Ancestral Knowledge in Agri-Allied Science

books for everyone
PAPERBACKS

NIPA® GENX ELECTRONIC RESOURCES & SOLUTIONS P. LTD.
New Delhi-110 034

www.nipaers.com
Browse, Search, Read & Buy...
Print Books
eChapters
eBooks
Publishers
Forthcoming
Subject
Catalogues
New Titles
Online Resources on:
✓ Current Affairs
✓ Reasoning, Logic & Aptitude
✓ Competitive Examinations
✓ English Language Lab
✓ Writing & Pronunciation Tools
✓ Personality Development
✓ Online Programmes for Professionals
✓ Interview Preparation
Pay Using
PhonePe
Google Pay
paytm
PayPal
UPI
CC Avenue

Ancestral Knowledge in Agri-Allied Science

Ratan Kumar Saha
Associate Professor & Head
Department of Fish Health & Environment

Dilip Nath
Research Associate (NAIP, Component III)
Department of Fish Health & Environment

Himadri Saha
Assistant Professor
Department of Fish Health & Environment

College of Fisheries
Central Agricultural University, Lembucherra – 799 210
Agartala, Tripura (West)

nipa® books for everyone PAPERBACKS

NIPA® GENX ELECTRONIC RESOURCES & SOLUTIONS P. LTD.
New Delhi-110 034

books for everyone
PAPERBACKS

NIPA® GENX ELECTRONIC RESOURCES & SOLUTIONS P. LTD.

101,103, Vikas Surya Plaza, CU Block
L.S.C. Market, Pitam Pura, New Delhi-110 034
Ph : +91 11 27341616, 27341717, 27341718
E-mail: newindiapublishingagency@gmail.com
www: www.nipabooks.com

For customer assistance, please contact
Phone: + 91-11-27 34 17 17
Fax: + 91-11- 27 34 16 16
E-Mail: feedbacks@nipabooks.com

ISBN: 978-81-19002-07-8

Composed and Designed by NIPA®.

डा. एस. अय्यप्पन
सचिव एवं महानिदेशक
Dr. S. AYYAPPAN
SECRETARY & DIRECTOR GENERAL

भारत सरकार
कृषि अनुसंधान और शिक्षा विभाग एवं
भारतीय कृषि अनुसंधान परिषद
कृषि मंत्रालय, कृषि भवन, नई दिल्ली 110 001
GOVERNMENT OF INDIA
DEPARTMENT OF AGRICULTURAL RESEARCH & EDUCATION
AND
INDIAN COUNCIL OF AGRICULTURAL RESEARCH
MINISTRY OF AGRICULTURE, KRISHI BHAVAN, NEW DELHI 110 001
Tel.: 23382629; 23386711 Fax: 91-11-23384773
E-mail: dg.icar@nic.in

Foreword

From ancient times, agriculture has been a major source of food for humanity and a provider of employment and economic benefits to those engaged in this activity. Since then, human-beings have evolved and subsequently developed technologies for meeting their livelihood through agricultural and allied activities. Those knowledge or technologies are synonymously called as Indigenous Technical knowledge or Indigenous Traditional Knowledge. These knowledge are very precious and in modernization era, these are getting lost. Those are needed to be preserved in the form of manuscript, technical paper or in digital form etc.

I am glad to note that Dr. R.K. Saha, Dr. D. Nath and Mr. H. Saha have compiled and edited all the articles received from different parts of India and published as a book entitled "**Ancestral Knowledge in Agri-Allied Science**". This book is an appreciable initiative towards preservation of ancestral knowledge in the agri-allied field. Several examples have been cited where theoretical and laboratory based technology failed due to location specific variability, technology developed based on ITKs showed significant acceptance.

I am confident that this book will prove helpful as a reference to students, entrepreneurs, researchers and others to improve their comprehension about the requirement by a locality and will help those rural actors to develop new technologies, using traditional knowledge along with imagination, creativity and determination to replace existing insignificant models.

Dated the 22nd April, 2013, New Delhi

(S. Ayyappan)

Preface

The book entitled "**Ancestral Knowledge in Agri-Allied Science**" is intended to be of vast application in the field of agriculture and it is hoped that it will prove of interest to numerous people working in the field of agri-allied sectors. The book is the compilation of several chapters contributed by different resource persons of the country working in the same field for several years.

This book is designed and covered several topics keeping in mind the background knowledge of the targeted reader of this book. The readers may be new in the field as well as the experienced professionals. The book has covered different Indigenous Traditional Knowledge (ITK) related to the agri-allied sector of the country. Another motto of this book is to document some of the knowledge of the indigenous people inhabiting in different geographical regions of the country with their own language, culture, tradition, belief, folklore, rites and rituals, which are eventually getting lost. Several researchers are working on ITK in different fields of agriculture and allied sectors. There is a demand on a standard book like this on the overview of ITKs of this sector.

A foreseeable criticism of this book is that it is patchy in its coverage of the all the ITKs of the country in the concerned field- some are covered extensively and some are not. However, efforts were being done to cover major topics of all sectors. This book is designed in such a way that will give an overview of ITK, the differences between Science & ITK, different tools and techniques used in ITK, classification, importance & utilization of ITK, and recent ongoing researches on ITK in different parts of India.

Another objective of this book is to encourage the study of ethnic knowledge on different field of agriculture and allied sector. It is also an attempt to circulate amongst a larger group of readers regarding the importance of indigenous knowledge in scientific world.

We hope this book will help all personals to gathered knowledge about the skills and techniques used by the ancestors in the field of agriculture and allied sectors for the benefits of mankind. Although we have complied that book is not primarily intended for the researchers and hope it will help them to concentrate their attentions where they are more needed.

In fact it is a matter of great pleasure for us to associate ourselves in the task for compilation of traditional information available with the farmers in the form of a book. We are very much indebted to Dr. S. Ayyappan, Secretary & Director General, Government of India, Department of Agricultural Research & Education and Indian Council of Agricultural Research, Ministry of Agriculture, New Delhi for providing us the foreword with his encouraging words. We are very much grateful to Dr. S.N. Puri, Vice Chancellor, Central Agricultural University, Imphal, Manipur for his kind support, cooperation and permitting us to publish this book for wider accessibility. The editors are highly indebted to Professor M. Rohini Kumar Singh, Director of Research, CAU, Imphal, Professor M. Premjit Singh, Director of Extension, CAU, Imphal and Professor J. R. Dhanze, Dean, College of Fisheries, CAU, Lembucherra, Tripura for their support, suggestions and guidance in compilation of this book.

The editors would like to express their gratefulness to all the contributors of different part of the country who have contributed their valuable studies and surveys as chapters and all of our CAU Faculty Members and Technical Staff Members for their timely help in preparation of this book. The authors are gratefully acknowledges the refereed scientists, teachers, researchers and friends for providing their help in writing this book to fulfill the students and researchers need. Last but not the least editors are sincerely acknowledges the help and cooperation of all of their family members and all the well wishers who have rendered their inspiration all along on completion of this book. Hope, this book will prove helpful to all those who are related with agriculture/ horticulture/ animal husbandry/ aquaculture/fisheries in one way or the other. Any error or omission is not intentional. Constructive scrutiny, suggestion and feedback are also welcome from all concerns and will be gratefully received and acknowledged for improving this book in all respect for the greater interest of this applied and fast growing field. It is our proud privilege and pleasure to express our deep sense of gratitude towards the publishers of this book for critically reviewing the book and publishing our compilations within short span of time. All feedback can be sent to feedbacks@nipabooks.com; or ratankumarsaha123@rediffmail.com; or ratankumarsaha123@gamil.com.

Editors

Contents

Contributors

Ratan K. Saha
Associate Professor & Head
Department of Fish Health & Environment
College of Fisheries, CAU
Lembucherra – 799 210
Agartala, Tripura (West)

Dilip Nath
Research Associate (NAIP, Component III)
Department of Fish Health & Environment
College of Fisheries, CAU
Lembucherra – 799 210
Agartala, Tripura (West)

Himadri Saha
Assistant Professor
Department of Fish Health & Environment
College of Fisheries, CAU
Lembucherra – 799 210
Agartala, Tripura (West)

A. K. Roy
Consultant (Statistics)
Department of Fisheries Economics and Statistics
College of Fisheries, CAU
Lembucherra – 799 210
Agartala, Tripura (West)

B. Saha
Assistant Professor
Department of Fisheries Extension
College of Fisheries, CAU
Lembucherra – 799 210
Agartala, Tripura (West)

T. Samajdar
Programme Coordinator
Krishi Vigyan Kendra, ICAR RC for NEH Region
Sangsanggre, Dobasipara, West Garo Hills
Meghalaya – 794 005

P. K. Pal
Assistant Professor (SS)
Uttar Banga Krishi Viswavidyalaya
P.O.Pundibari, Cooch Behar
West Bengal – 736 165

D. M. Firake
Scientist
Division of Entomology
ICAR Research Complex for NEH Region
Umiam, Meghalaya – 793 103

P. D. Firake
Scientist
Division of NRM
ICAR Research Complex for NEH Region
Umiam, Meghalaya – 793 103

G. T. Behere
Scientist
Division of Entomology
ICAR Research Complex for NEH region
Umiam, Meghalaya – 793 103

N.S. Azad Thakur
Scientist
Division of Entomology
ICAR Research Complex for NEH Region
Umiam, Meghalaya– 793 103

Arunava Samanta
Associate Professor (Research)
AICRP on Sub-Tropical Fruits
Bidhan Chandra Krishi Viswavidyalaya
Kalyani, Nadia – 741 235
Wes Bengal

Archan Kanti Das
Principal Scientist
Central Inland Fisheries Research Institute
ICAR, Barrackpore
Kolkata – 700 120
West Bengal

Sandip Patra
Scientist (Agril. Entomology)
Division of Crop Improvement
ICAR Research Complex for NEH Region
Umiam – 793 103
Meghalaya

M. K. Datta
Assistant Professor (SS)
Department of Aquaculture
College of Fisheries, CAU
Lembucherra – 799 210
Agartala, Tripura (West)

S. C. Mandal
Assistant Professor
Department of Aquaculture
College of Fisheries, CAU
Lembucherra – 799 210
Agartala, Tripura (West)

Pampa Bhattacharjee
Assistant Professor (SS)
Department of Fisheries Resource Mabnagement
College of Fisheries, CAU
Lembucherra – 799 210
Agartala, Tripura (West)

Debtanu Barman
Project Assistant (COE, DBT)
Department of Fisheries Resource Management
College of Fisheries, CAU
Lembucherra – 799 210
Agartala, Tripura (West)

A. Chakraborty
Scientist
AICRP on Potato
Directorate of Research, BCKV
Kalyani, Nadia, West Bengal – 741 235

Subham Dey
PhD Scholar
Department of Plant Pathology
BCKV, Mohanpur, Nadia
West Bengal – 741 252

A.U. Muzaddadi
Assistant Professor (SS)
Department of Fish Processing Technology
College of Fisheries
Central Agricultural University
Lembucherra, Tripura – 799 210

Ambalika Ghosh
Assistant Director of Fisheries, Howrah
Govt. of West Bengal
21, Rose Mary Lane, Howrah – 711 101

Partha Pratim Bhattacharjee
Senior Research Fellow (ICAR)
Ecology and Biodiversity Laboratories
Department of Zoology
Tripura University, Suryamaninagar
Tripura – 799 022

B. K. Agarwala
Professor
Ecology and Biodiversity Laboratories
Department of Zoology
Tripura University, Suryamaninagar
Tripura – 799 022

Rizwana
Assistant Professor
Faculty of Fisheries, Shuhama, Ganderbal
Sher-e-Kashmir University of Agriculture
Sciences of Technology
Kashmir

T. H. Bhat
Assistant Professor
Faculty of Fisheries, Shuhama, Ganderbal
Sher-e-Kashmir University of Agriculture
Sciences of Technology, Kashmir
Jammu and Kashmir

M. H. Balkhi
Professor
Faculty of Fisheries, Shuhama, Ganderbal
Sher-e-Kashmir University of Agriculture
Sciences of Technology
Kashmir (J.&K.)

Poonam Sharma
Assistant Professor
Faculty of Fisheries, Shuhama, Ganderbal
Sher-e-Kashmir University of Agriculture
Sciences of Technology
Kashmir (J.&K.)

T. Shantibala
Scientist
Animal Resources Division
Institute of Bioresources and Sustainable Development
Department of Biotechnology
Govt. of India
Takyelpat, Imphal – 795 001, Manipur

R. K. Lokeshwari
Scientist
Animal Resources Division
Institute of Bioresources and Sustainable Development
Department of Biotechnology
Govt. of India
Takyelpat, Imphal – 795 001
Manipur

Gusheinzed Waikhom
Scientist B
Animal Resources Division
Institute of Bioresources and Sustainable Development
Department of Biotechnology
Govt. of India
Takyelpat, Imphal – 795 001
Manipur

J. Majumder
Senior Research Fellow (ICAR)
Ecology and Biodiversity Laboratories
Department of Zoology
Tripura University, Suryamaninagar
Tripura – 799 022

Pabitra H. Patra
Assistant Professor
Department of Pharmacology & Toxicology
College of Veterinary Sciences and Animal Husbandry
Govt. of Tripura, R.K. Nagar
Agartala – 799 008, Tripura

Avijit Haldar
Senior Scientist
Animal Reproduction Division
ICAR Research Complex for North Eastern Hill Region
Tripura Centre, Agartala
Lembucherra – 799 210
West Tripura

Chintu Deb Barma
Assistant Professor
Department of Veterinary Extension
College of Veterinary Science and Animal Husbandry
Govt. of Tripura, R.K. Nagar, Agartala
West Tripura

Rajkumari Sanjukta
Scientist
Division of Animal Health
ICAR Research Complex for North Eastern Hill Region
Umroi Road, Umiam, Meghalaya – 793 103

Debasish Behera
Assistant Professor
Department of Veterinary Pathology
College of Veterinary Sciences and Animal Husbandry
Govt. of Tripura, R.K. Nagar
Agartala – 799008
Tripura (West)

Anil Datt Upadhyay
Department of Fisheries Economics and Statistics
College of Fisheries
Central Agricultural University
Lembucherra – 799 210
Tripura

Pranabjit Bardhan Roy
Assistant Professor
Department of Environmental Science
Govt. Degree College, Kamalpur
Dhalai – 799 285
Tripura

W. Ramdas Singh
Assistant Professor
Department of Pharmacology & Toxicology
College of Veterinary Sciences and Animal Husbandry
Govt. of Tripura, R.K. Nagar
Agartala-799008
Tripura (West)

Hijam S. Devi
Ph.D Scholar
Department of Entomology
Punjab Agricultural University (PAU)
Ludhiana – 141004, Punjab

Krishna Prasad Sharma
Department of Fisheries Extension
College of Fisheries, Central Agricultural University
Lembucherra – 799 210
Agartala, Tripura (West)

R. Lodh
Research Scholar, UGC
Ecology and Biodiversity Laboratories
Department of Zoology
Tripura University, Suryamaninagar
Tripura – 799 022

Ramanuj Chakraborty
Subject Matter Specialist (Fish)
Krishi Vigyan Kendra
Department of Agriculture
Govt. of Tripura
Salema R.D. Block, Kamalpur – 799 278
Dhalai District, Tripura

Subrata Shib
Programme Coordinator
Krishi Vigyan Kendra
Department of Agriculture
Govt. of Tripura, Salema R.D. Block
Kamalpur – 799 278
Dhalai District, Tripura

H. Bharati
PG Student
Central Institute of Fisheries Education
ICAR, Mumbai, Maharashtra

Yumlembam Jackie Singh
Assistant Professor
Department of Fisheries Extension
College of Fisheries
Central Agricultural University
Lembucherra – 799 210
Agartala, Tripura (West)

Rajita Devi
Department of Fisheries Extension
College of Fisheries, CAU
Lembucherra – 799 210
Agartala, Tripura (West)

Olik Jomang
Department of Fisheries Extension
College of Fisheries, Central Agricultural University
Lembucherra – 799 210
Agartala, Tripura (West)

Pema Chotten Wangja
Department of Fisheries Extension
College of Fisheries
Central Agricultural University
Lembucherra – 799 210
Agartala, Tripura (West)

Chapter – 1

An Overview of Indigenous Technical Knowledge in Relation to Agri-allied Sector

R.K. Saha, D. Nath and H. Saha

INTRODUCTION

Indigenous knowledge is the knowledge of the indigenous people inhabiting different geographical regions of the world with their own language, culture, tradition, belief, folklore, rites and rituals. In course of their close interactions with nature and natural resources, they are to make a certain decisions as to solve their problems they encounter in their day to day life while managing the land and environmental resources for survival.

Several researchers explained traditional knowledge as holistic, encompassing many fields of scientific knowledge, such as climate, hydrology, soils, plants and animals (Howes and Chambers, 1980, Jungerius, 1985, Wilken, 1987, Agrawal, 1995). Therefore, indigenous knowledge so developed is based on necessities, instinct, curiosity and observations of ethnic groups to mitigate the immediate problems. Several logical observation was used by them in the decision making process as to how, when and where to act depending on the situations. This knowledge has been tested using the thumb rule of trial and error methods over a period of time through generations and validated to make the established knowledge for the purpose for which it was designed for use of the ethnic group who propounded for the first time. Thus, indigenous knowledge varies from tribe to tribe and usually passes on to the next generations though the words of mouth, actions or even practices, usually by the elders of the family/society.

CONSTRAINS AND REMEDIES

These ethnic based cumulative knowledge took generations of time to penetrate into the social fabrics of inter and intra- ethnic groups because of communications gap and orthodox nature of the society. Eventually, this local knowledge in course of time gets socially accepted and validated which finally inters into the social life and subsequently become the indigenous traditional knowledge (ITK) of the society as a whole.

In regard to the transfer of knowledge/technology, the early approaches were and continue to be, characterized by a top down approach and a transfer of technologies from the western world. The local people were seen as part of the problem to be solved despite the fact that those people survived with their approaches for centuries. Technologies developed in areas with totally different biophysical conditions were imposed on the people.

In some cases these technologies failed, in other cases they met with a lot of resistance to the extent that more effort was put in convincing the local people to adopt them and in yet others they worked but for a short time after which the land resources base was destroyed. The people either had to move away or spend almost all available resources to at least produce enough for their survival.

We in the developing world were unaware that the any measures introduced were based on the practices applied in the west. We apply them to our fragile ecosystems with no regard for ecological concepts and continue to witness failure. Often, farmers are handicapped because they lack a scientific knowledge on land use pattern, water parameter, pest and disease symptoms, ecology, natural enemies, and development pattern of crops, appropriate control measures, soil condition and its effect of weather conditions; which ultimately affect their productivity.

So there is much to do. It will continue to be very difficult to solve all such problems as long as the local community's ways are not understood. After all it is the local people who manage the natural resources and who will continue to make decisions on what should be done. Until bureaucrats and policy makers faithfully represent and apply what is on the ground, not let a few academicians dictate what is on the ground but let the people on it to say how they do things, until they reflect what the local people want and more importantly until they reflect the knowledge that comes from the grass roots, we will not be able to make sustainable management a reality.

A new approach is needed if this problem is to be solved. This is an approach based on the ecological and socio-economic understanding of the environment and the local people and their relationship. A new approach of natural resource management is emerging. Cardinal to the new approach is the prerequisite for active participation and collaboration between the local users and the technicians and development agents. Of direct relevance to this new paradigm is the increase recognition of the importance of Indigenous Technical Knowledge (ITK).

Despite its potentiality in addressing the environment issues including agriculture, indigenous traditional knowledge could not develop to its fullest extent primarily for fast technological development and secondly due to crude and outdated nature of the knowledge, that too in the process of degeneration with the passing away of elders and finally due to the exposure of the society to their modern innovations. As the elderly persons in any ethnic group hold the key and custodians of the traditional knowledge, they are the masters to make correct decisions to overcome the adverse situations of their immediate environment. Being a local need based knowledge, indigenous traditional knowledge system is based on local resources without any external inputs. Eco-friendly manipulation of their immediate environment, judicious application of plant and animal products either in raw or simple processed forms are important components of indigenous knowledge system. Land being the basis of survival of mankind, indigenous knowledge system of indigenous people revolved mostly about the ecological management of their land and environment in harmony with nature. Here lies the importance and significance of indigenous traditional knowledge and therefore it should be protected, preserved and documented in the interest of future generations before it is lost.

DIFFERENCES BETWEEN TRADITIONAL AND SCIENTIFIC KNOWLEDGE SYSTEM

The scientific community believes that there is scientific knowledge works behind traditional practices in agriculture, pest management, animal husbandry, health management, disease control etc. However, underlying differences are also observed among the scientists in this regard. Scientists working in the field of applied agriculture with experience in the area of Indigenous Plant Protection Practices (IPPPs) believe in the science behind farmers' practices. Similarly, scientists with specialization in Vedic agriculture and taken up teaching assignment also feel the same, even though they have never carried out any work on IPPPs.

Vedic agriculture provides a platform to explain modern science from the perspective of indigenous/cultural science.

The socially validated indigenous knowledge which shares common values and gains popularity with human element attached to it particularly the culture of the ethnic groups may be known by the name indigenous traditional knowledge. However, now a days it is known to people synonymously by various modern names such as indigenous technical knowledge, technical knowhow of indigenous people, peoples knowledge etc. (Singh and Sureja, 2008).

The second group of scientists with exposure to the traditional plant protection methods and those who are confined to laboratory research or basic research realize the role of science behind local agricultural practices quite differently. They differentiate traditional science from the modern science as alternative forms. They state that science in traditional knowledge exists in a crude form as it is integrated with the socio-cultural and economic systems of the society; therefore difficult to distinguish. Here, science is understood in terms of its level of refinement. The statement can be well understood from the experience cited by a scientist. Scientists understand the reason of science from the practitioners' perspective, which, cannot be experimented by detaching the religious functionalities. Thus, science in its crude form has so far not been experimented outside the practitioners' society. The third group of scientists working in the area of IPPPs with reference to the research in both laboratory and field types, hold a similar kind of view. The case below would provide an explanation to the fact.

There are some scientists who opine 'not all the Indigenous Technical Knowledge (ITKs) can be testified in the laboratory condition, as the laboratories do not have sophisticated equipments to evaluate the exact scientific rationality of ITKs'. ITKs have scientific viability; however, the laboratory condition may not provide a conducive environment to prove the science behind ITK. Further, they added, the science behind ITKs can be termed as coincidence between the belief and effect and could be verified only through observation.

In the effort to justify the prevailing belief system behind certain IK, it is found that coincidence happens, and becomes the tool to prove certain ITKs as effective, if not scientific. Sometimes new technology is also anchored on the traditional belief system for gaining acceptance among the community people. A scientist cites his experience of how traditional belief systems help in the promotion of new technology. The scientific community believes in the concept of 'peasant science', i.e., embedded in indigenous practices. Scientists consider any practice as science from a methodological standpoint.

The written reference of indigenous practices is a source for asserting them as science, but they cannot be considered scientific unless brought into the laboratory. Documentation of any local agricultural practice helps in the dissemination of the knowledge to the non-practitioner group, its further use in both on-farm and laboratory. Therefore, scientific elements cannot be manifested through practices, but they are sources of science. Here, science is located in the practitioners' perspective therefore it is context specific.

A dichotomy between the ITK and scientific knowledge systems may be made on the following grounds:

1. **Substantive grounds -** because of differences in the subject matter and characteristics of traditional and scientific knowledge.
2. **Methodological and epistemological grounds** - because the two forms of knowledge employ different methods to investigate reality.
3. **Contextual grounds** - because traditional knowledge is more deeply rooted in its environment.

This is not to say that ITK does not change, but rather than those changes which occur are in nearly all instances comparable to the achievements of what Kuhn (1962) termed it as 'normal science', or to the detailed working out of relatively minor 'puzzles' within an established 'paradigm' of thought. Science, in contrast, constantly carries with it the possibility of 'revolutionary change' in which one paradigm would be destroyed and replaced by another. Different scientist has contrasted and evaluated "Science and ITK" based on three criteria:

i. As systems of classification;
ii. As systems of explanation and prediction;
iii. In terms of speed of accumulation.

While ITK and science are comparable on the first criterion, science is generally superior on the second and markedly superior on the third.

On the basis of above interference we can summarize the difference between science and ITK are as follows:

	Science		ITK
a)	The mode of science has a greater ability to break down data presented to the senses and to reassemble it in different ways.	a)	The mode of ITK is 'concrete' and relies almost exclusively on intuition and evidence directly available to the senses.
b)	It is an open system whose adherents are always aware of the possibility of alternative perspectives to those adopted at any particular point of time.	b)	It is a closed system, is characterized by a lack of awareness that there may be other ways of regarding the world.
c)	Human life is generally regarded as superior, with a moral right to control other life forms.	c)	All parts of the natural world are regarded as animate, all life forms as interdependent.
d)	Knowledge is transmitted largely through written media.	d)	Knowledge is transmitted largely through oral media
e)	Knowledge is generally learned in a situation, which is remote from its applied context.	e)	Knowledge is developed and acquired through observation and practical experience.
f)	Knowledge is essentially reductionist, quantitative, analytical and theoretical.	f)	Knowledge is holistic, intuitive, qualitative and practical.
g)	Knowledge is generated largely specialist researchers on a synchronistic (short term) time scale.	g)	Knowledge is generated by resource users in a diachronic (long term) time scale.
h)	The nature and status of particular knowledge is influenced by peer review, and is held by individual specialists.	h)	The nature and status of particular knowledge is influenced by socio-cultural factors such as spiritual beliefs, and is communally held.
i)	Knowledge is used to make suitable decisions under variable conditions.	i)	Knowledge is used to put forward hypothesis and to verify underlying laws and constants.
j)	Mental models, such as schemata, paradigms, perspectives, beliefs and viewpoints, help individuals to perceive and define their world.	j)	Explanation behind perceived phenomena is often spiritually based on subjective.
k)	The cognitive elements of ITK center on what Johnson-Laird (1983) calls "mental models", where human beings create working models of the world by making and manipulating analogies in their minds.	k)	Explanation behind perceived phenomena is essentially rational and objective.

On the basis of above mentioned discussions, we can summarize ITK is distinct from scientific knowledge in the following respects:

i. Localness of ITK – derivation from close, long relationships between people and a specific land / resource area gives ITK its focus.
ii. Ownership by the local community, even though non-homogeneously, integrates ITK with social priorities.
iii. Classifications in ITK are usually assumed to be more based on functionality or purposiveness than those of Scientific Knowledge.
iv. ITK is more holistic because of the purposiveness – e.g. an emphasis on holistic, combinatorial explanations. Although this holism is usually seen in opposition to reductionist epistemology, holistic systems approaches are also utilized in science.

Nevertheless, ITK has more elements in common with scientific knowledge, which are as:

i. Dynamism - Interest and ability to incorporate new knowledge from other (outside) sources, notwithstanding they may contradict held beliefs. (Brouwer, 1995).
ii. The imperative for taxonomies, as the building blocks of some kind of explanations.
iii. Identification of specific conditions under which general 'laws' or explanations will hold.
iv. Drive to experiment – to explore the effects of altering the underlying conditions; (for ITK, see Richards, 1989).
v. Knowledge is unevenly distributed within a community of experts - whether the 'controlling' ITK arrogated by male elders/ castes (McCall, 1995; McKinnon, 2001), or, by closed scientific 'academies'.
vi. ITK is often devalued as "culturally-embedded, encoded in ritual"; but there is a parallel in growth-driven, individualistic, materialist cultures that form the context of western science funding.

'ITK' LINKED RESEARCHES ON AGRI-ALLIED SECTORS

There are different projects have been operated by the scientists of agri-allied sectors. The nature of projects include 'protection of tomato seeds using lemon juice', 'use of cow dung for sprouting of Desi Alu (local potato variety)', 'management of soil structure', 'use of botanicals (straw and coconut leaf) for protection of crop plants', 'safe pesticide trials', 'escaping drought through application of both chemical and bio-fertilizer', 'plant protection with herbal products', All India coordinated pulses improvement project: preservation of pulses seeds with Polang (a tree found locally), mango, Suniari (a local creeper) leaves is mixed with ash to preserve pulses seeds in straw packets', 'bio-pesticide and insecticide testing' and so on. Research on bio-pesticide and preparing safe insecticide are mainly carried out by corporate funding agencies such as Biotechnology Nagarjuna International, BASF India Ltd., Singonta, Rhone Poulemic India Ltd. Some important fields of ITK related to agri- allied science are furnished below.

i. Indigenous Technical Knowledge (ITK) in Plant protection and Pest management

Insect pest cause heavy losses to crops especially the production of vegetables. Curative control measures are practiced to protect them. Among them, synthetic insecticides have been used since long time, but have serious drawbacks (Sharaby, 1988). These insecticides carry potential health hazards and affect non-target organisms and environment (Islam et al., 2003). The traditional pest control practices play an important role in the management of agriculture and it is an inevitable practice for sustainable agriculture. The proper control of pests minimizes economic losses and damage to the environment (Murdoch et al., 1985).

Scientists have taken up work on indigenous plant protection methods specifically, 'neem oil based pest control, vegetable pest management, soil testing, soil classification, use of local plants and leaves for seed preservation, testing of local rice pest resistant variety, and experimentation of local protection methods like wood ash, charcoal ash, and grinded snail in vegetable and rice cultivations'. Perhaps, the scientists feel that these above areas provide a wider scope to work on cultural indigenous plant protection research. The agricultural scientists, who were in charge of regional offices of Orissa University of Agriculture and Technology (OUAT), have acquired fair knowledge of the various indigenous practices of plant protection. Scientists from the department of plant pathology, nematode and meteorology are least exposed to research on IPPPs. Nine species of non-sting bees (Meliponinae) were discovered thanks to the Kayapó Indian's ethnoentomology (in the Amazon basin) (Posey, 1986).

The IPM strategies developed and promoted by the Governments are now based on these traditional practices for better management. If more attention is given towards the production or promotion of ITK based products on cottage scale, it could be an economically viable option for sustainable development of eco-friendly pesticides/insecticides. Moreover, due to rapid urbanization, the ethnic groups are promptly reducing their conservative attitude for which the long earned ITK that they are carrying from generation to generation are undocumented and quickly disappearing (Deka et al., 2006). In this context, collection, compilation and scientific evaluation of the ITKs are very important in terms to strengthen modern agricultural research and biodiversity conservation.

Several workers reported tobacco, turmeric, chilly and lime water have traditional usage in India (Deka et al., 2006; Pathak et al., 2001) and abroad (Anonymous, 1977) for pest control. Nath and Ray (2012) reported that about 77.5% farmer Cachar district of Assam practices the traditional methods of pest management. The farmers of this district use tobacco and turmeric as traditional practices for pest control. Wood ash, cow dung, chilly, turmeric, tobacco all of these are biological origin where lime is only product which is not originated from biological world (Nath and Ray, 2012). The most used biological origin of traditional practices includes neem, pumalo (*Citrus grandis*), phutuka (*Melastoma malabathricum*), drumstick (*Moringea oliefera*), fern, bamboo, duck and cow dung etc. (Deka et al., 2006). However, only few components, such as kerosene and tyre are derived from synthetic sources.

The traditional practices, if properly incorporated in pest management strategies, can help to a great extent to check an increasing pest problems because these practices are eco-friendly as well as sustainable manner of pest management (Nath and Ray, 2012).

ii. Indigenous Technical Knowledge (ITK) in land use pattern

There are pernicious and widespread problems of land degradation in most of the tropics and subtropics. The problem is made even worse as the fragile tropical ecosystems absorb a growing population (El Ahry, 1988). A new approach to land husbandry is emerging. Cardinal to the new approach is the prerequisite for active participation and collaboration between the local users and the technicians and development agents. Of direct relevance to this new paradigm is the increase recognition of the importance of Indigenous Technical Knowledge (ITK).

The people of Central Northern Zambia practice an indigenous farming system called Chitemene, which in essence is a type of shifting cultivation. In this practice the trees are lopped and pollarded and the cut wood piled and burnt on a selected site and crops planted in the ash-covered area. The trees are cut from a lager area (outfield) they are burnt in the center (infield) which is used for crop cultivation while the cleared land is left to regenerate (wgbis.ces.iisc.ernet.in/energy/HC270799/LM/SUSLUP/.../76.pdf). A new

Chitemene is opened each year while the opened Chitemene is used for a period of 5 to 6 years after which it is left to regenerate for 10 to 30 years. Important ITK in this system include:

a) selection of an appropriate area for tree cutting and burning which will not disturb the land resources;

b) the way each tree species is cut and piled and burnt. While some trees are pollarded, others are merely pruned and yet others are coppiced;

c) the timing when to clean the land and when to burn to avoid damage to the resources;

d) the method of burning to get the best ashes;

e) the selection of which crops to start with and which crops to follow in subsequent years;

f) the farmers also decide which crops are intercropped;

g) finally the understanding and identification of indicators showing that the piece of land is exhausted and needs resting.

The Chitemene system was a very viable system which is now failing because of the increased population and consequent lack of land which prelude that the used land parcels regenerates and recovers. A system grafted onto this indigenous farming practice had to be developed. Improved fallow was tried through the Soil Conservation and Agroforestry Extension Programme (SCAFE) as one method to improve fertility. In improved fallow the farmer merely plants the land which is not under cultivation for different reasons with some trees or shrubs. The planted tree species are selected to fit the farmers labour and other socio economic aspects and indeed the farming methods the farmers apply.

The development of the improved fallow as a type of land management system fitted very well with the indigenous system. The improved fallow is different from the indigenous farming system only in the sense that the land is made to regenerate faster than it does naturally by planting fast growing tree species and preferably nitrogen fixing tree species. This is a practice, which in Central Zambia is purely based on and developed from ITK.

iii. ITK in fish farming

Indigenous technical knowledge in the field of fish farming is a practical knowledge of the local farmers (Saha and Nath, 2013). Unlike the modern techniques, The ITK had a significant level of adoption either knowingly or unknowingly, among the substantial part of the resources in NE region comprises capture of fisheries, the management and identification of ITK in fish culture activities. Selection of fishing methods and gear are influenced by various factors such as physiography of the water body, nature of fish stock, characteristics of the raw material from which gear are fabricated and standard of living (Choudhury, 1992).

There are different ethnic fisherman communities in Dhalai District of Tripura who are practicing technical knowledge in their day to day life (Fig. 1). They are practicing integrated fish based farming system where the integration of plants and animals they have incorporated. These include *Fish-cum-vegetable (bottle gourd) culture, Paddy-cum-fish culture and Fish-cum-duckery.* They are also practicing the method of *protection of fishes from enemies* where the use of bamboo and tree twigs in the pond which protects the fishes from fish eating animals (enemies), from pouching and also provides the foods grown to the underwater parts of the twigs to the fishes. The farmers also invented the method of *keeping the freshness of bamboo* for which they are using ponds water for the protection of bamboo from damage. Regarding fishing methods the *Garo* community has made bamboo based different *fishing traps* which are locally known as *usha* (Saha and Nath, 2013).

Kalita et al. (2010) reported different ITKs on fish harvesting of Karbi-Anglong district of Assam. They documented some unique indigenous knowledge related to fish harvesting. These includes fish harvesting, fishing with attractant, piscicidal plants, community fishing, *matia chali, kalpotua chali*, pot fishing and crab fishing. Under plant piscicides, *smart weeds, karoi tree* and *moin* were documented. Under community fishing, scare fishing and *dalbandhi maach mara* were documented by them.

The Mizo tribes of north east India developed and practiced different types of fishing methods since time immemorial. Some methods can be considered as traditional as well as non-traditional since they are well known in some other parts of the country with some variation. The people of Mizoram used locally available materials and apply indigenous ideas and skills (indigenous technical knowledge, ITK) for fishing. Various fishing techniques depend on various behavioural pattern and microhabitat type of fishes. Thirteen species of plants and plant products are used for fishing (Lalthanzara and Lalthanpuii, 2009). The success of these fishing techniques depends on various factors like selection of site, time, efficiency of materials used and availability of fish, etc. For successful fishing some attractants or lure is popularly employed in other parts of the country (Dutta and Bhattacharjya, 2009; Srivastava et al., 2002). Some sub-tribe of Mizo like Pawi, Hmar, Lai, Mara, etc., have their own methods of fishing which are usually specially modified methods of the existing one, e.g., nghahringdawh by Pawi (Lalthanzara and Lalthanpuii, 2009).

Bottle gourd cultivation along with fish

Duba (natural)

Duba (Natural)

Duba (man-made)

Integrated fish culture

Use bamboo and twigs in the pond

Bamboo made fish trap (Conical)

Bamboo made fishing tap (Cylindrical)

Bamboo made fishing trap (basket type)

Fig. 1: Fishermen communities in Dhalai District of Tripura practicing different technical knowledge (*Source*: Saha and Nath, 2013).

iv. ITK in Ethnobiology and Ethnomedicine

In the last years, traditional populations' wisdom has been deemed as extremely important for the development of studies in several areas of natural sciences. This knowledge is the aim of ethnoscience studies, where the prefix ethno refers to the system of knowledge and cognition that are typical of each culture (Campos, 2002). Among the ethnosciences, the ethnobiology deserves special attention by involving the analysis of the classification system on nature and for having a deep link with the subjects of botany, zoology, and ecology.

The traditional communities in the world over are the treasure trove of accumulated knowledge and wisdom particularly about their natural resources. In recent years, the plants used traditionally for curative purpose, the traditional knowledge system handed over from generation to generation by traditional communities by oral method is still continuing in many developing countries, have attracted attention of the research areas (Kargioglu et al., 2008; Simsek et al., 2004; Ozgen et al., 2004). Most of the plants compounds employed in modern medicine were first discovered through Ethnobotanical investigations. There are some 130 plants derived compounds which currently used in western medicine and 74% of these have been discovered through follow up research work to verify the authenticity of the information concerning the folk/ ethnomedical use of plants (Farnsworth, 1988).

The medicinal properties of a plant are due to the presence of certain chemical constituents. These chemical constituents, responsible for the specific physiological action, in the plant, have in many cases been isolated, purified and identified as definite chemical compounds. Quite a large number of plants are known to be of medicinal use remains uninvestigated and this is particularly the case with the Indian flora. Furthermore, a growing world-wide interest in the use of phytopharmaceuticals as complementary or alternative medicine, either to prevent or to ameliorate many diseases, has been noted in recent years. Therefore, documentation of the indigenous knowledge through ethnobotanical studies is important for both *viz*., conservation and initialization of biological resources (Muthu et al., 2006). Several workers of the country reported different plant products for medicinal uses (Rajendran et al., 2008; Sen and Behera, 2008; Sen et al., 2008; Arya and Agarwal 2008; Das et al., 2008) and abroad (Ahmad et al., 2008; Omobuwajo et al., 2008) which are used in the treatment of various diseases (Fig. 2).

Shri. Dhurve (medicine man) of Goregaon

Shri. Pandhre (medicine man) of Deori

Costus speciosus (Koenig) Sm

Gloriosa superba L.

Fig. 2: Ethno-botanical observations of some important plants from Goregaon and Deori clusters of Gondia district (*Source*: Rahul et al., 2011)

The formulation of Ayurvedic and Unani medicine also have been achieved by following modern scientific quality control procedures both for the raw material and the finished product. The phytochemical constituents found to be present in the raw material used for the preparation of *Yogaraja guggulu* possibly facilitate the desirable therapeutic efficacy of the medicinal formulation as a whole in ailments, and also could help in knowing the underlying mechanisms of pharmacological action (Simha et al., 2008).

The importance of these traditional medicines has been realized world wide as some of them proved to be very effective and some other prescriptions of these traditional healers may be of benefits to human kind when thorough scientific analysis is conducted into their properties. At the same time the complications caused by some of the medicines prescribed by the traditional healers should also be taken into considerations. Herbal therapy is not only cost effective but also provides means for the treatment of many diseases, which are considered to be incurable in other system of medicines. It is thus important that modern scientific studies be done on these medicinal plants so that the plants may be used as remedies in a more rational and scientific manner. In this way such ethnobotanical studies enable the transfer of knowledge on plant based treatment (our natural inheritance) to the future generation (Rahul et al., 2011).

The documentation of the ethnic uses of plant resources is very important for a variety of reasons. The basic data, provided for economically important plants should encourage further studies aiming at their large scale cultivation and economic welfare. The facts and figures in case of ethnomedicinally important plants should be pursued for further studies including chemical and pharmacognostic analysis (Rai and Bhujel, 2002).

For each plant species, the information lists plant name, crude drug preparation, and the method of use. Such documentation not only provides opportunities, but also holds potential for developing products for the pharmaceutical sector, safeguard from bio-piracy and above all sustainable use.

v. ITK in agricultural technology

To improve farmers' production, researchers have developed some agricultural technology, which however produces several consequences, in the usage by farmers. Oladele et al. (2008) reported the farmer's perception of the consequences of arable crops technology in Oyo state. They discovered the benefits, adoption, social participation and contact with extension agents.

vi. ITK in natural resource management

Informal rural social institutions and indigenous knowledge systems as social capitals have often played significant role in conserving, managing and sustaining the indigenous biodiversity and natural resource management. Singh and Sureja (2008) documented the resource of poor farmers' informal social institutions and common practices for conservation and management of indigenous sugarcane varieties in Azamgarh district of Uttar Pradesh. The data were collected from the resource-poor and resource-rich farmers by using the focus group discussions and personal interview methods. They reported that farmers have developed an informal institution and a cumulative body of indigenous knowledge based on trial and error to conserve, adapt and manage the indigenous sugarcane varieties. The varieties and adopted approaches are found to be compatible with the socio-economic and biophysical indicators of rice-wheat based agroecosystem. Cultural feasibility and local resources available with farmers are basic attributes in facilitating the chain of cooperation through *Sajha-Kolhad* to help each other for management of indigenous sugarcane varieties in holistic manner. The informal institution and knowledge involved in managing the indigenous sugarcane varieties were found to be environmentally sound, economically viable, socially justifiable and more humane in nature. With regard to production stability and agronomic and cultural attributes, indigenous sugarcane varieties are more compatible than the improved varieties (Singh and Sureja, 2008).

Murugan et al., 2008 worked on acrid groves (*kavukal*) of Kerala. The vegetation in the groves is highly varied viz. mangroves, fresh water swamps, or other tropical forest types. Deities worshipped in the groves are also highly varied. One such type is the sacred grove dedicated to serpent God and serpent worship is considered to be one of the oldest and most prevalent forms of nature worship in the world. While serpent worship is seen through out India, only in Kerala, people worship serpent in the sacred groves dedicated to them called, *Sarpa kavu*. The study brought out three broad types of groves in Palakkad, *viz.,* the primitive, the recent and the sacraments devoid of groves. Worship, rites and rituals associated with the grove include both primitive ones like *Noorum palum*, *Kalemezhuthupattu* and the *Vedic* types like *Sarpa bali* and *Payasa homam*. They reported the existence of groves devoid of any vegetation, indicating that sacred groves can be relicts from a past socio-cultural epoch, which served to transmit the cultural heritage generations from pre-historic time. Conservation of natural resources in the past involved many *taboos*, rituals and other religious practices and sacred groves was such a traditional socio-cultural mechanism aiming at nature conservation that integrated socio-cultural aspects for conservation.

Researchers also worked on Indigenous Technical Knowledge (ITK) relating to soil and water management systems suitable for different agro-climatic conditions of India. Lal and Verma (2008) reported the ITK regarding soil and water management systems in the remote and rural area of Himachal Pradesh. They studied in several parts of the Himachal Himalaya villages which are located at the foothills or in the valleys. Villagers often harvest rainwater by building small water storage ponds commonly called *chal*. In hills, there is traditional practice to plough the fields early in the morning before dew or fog is evaporated for the conservation of water. Spring water is collected in small reservoirs scattered at intervals on the high uplands and then drawing water from these ponds when required. Heavy snow is melted with the help of ash. Ash is spread / broadcasted over field for weed control, surface soil crust breaking, and making soil more porous by conserving the rainwater. In Kinnaur district, soil mulching is done with *Pinus*, *Chilgoza* needles and grasses from the *kandas*/hill top. Mulching conserves soil moisture in the field and also helps in maintaining the soil temperature. In the lower areas of Himachal Pradesh during the rainy season, roof water is collected in dugout structures known as *diggi* in Kangra and *khatis* in Hamirpur and Bilaspur district. *Khatis* provides stored water, which is the main source of water. Farmers have been managing soil and water by making arrangements for the safe disposal of excess water.

The housing practices and human settlement strategy in civilization became the symbol of protection and safety for human existence. Whenever people think of housing construction, ecological factors have been given the principal contemplation to cope with the hostile natural calamities and unusual environmental behaviour. People around the world still maintain and practice this trend historical, pragmatic and situational in housing construction. Indigenous people across the world have hereditarily been exercising this sort of housing technology for years that includes ecological and environmental reflection. In north-eastern part of India the tribal people make their house in such ecological manner. They don't disturb the landscape and natural resources. They use the common available natural resources *viz*., wood, sun grass, bamboo, and cane etc for construction of house. The common name of such houses is known as *tonghar* which is constructed above the bamboo or wooden platform (Fig. 3).

Fig. 3: Tonghar of Halam community at Dhalai district of Tripura.

Whenever people think of housing construction, ecological factors have been given the principal contemplation to cope with the hostile natural calamities and unusual environmental behaviour. People around the world still maintain and practice this trend historical, pragmatic and situational in housing construction. Indigenous people across the world have hereditarily been exercising this sort of housing technology for years that includes ecological and environmental reflection. However, even modern architectural design embodies the discourse of 'environmentally-sound' in construction process. With the increased economic dimension of human life, economic factors are also reflected in the domain of housing technology that includes cost-effectiveness and sustainability, etc. Nasir Uddin, (2008) reported such traditional housing technologies, mud-made housing technology bring widely practiced in different parts of Bangladesh. It explores the state and architecture, ecological and economic reasoning of the mud-made housing technology of Bangladesh.

CONCLUSION

Some of the Indigenous practices used by different communities of India and abroad has discussed above. But these are not enough. There are several ethnic communities in the world are practicing their own technology in agri-allied sectors. Such technologies are very important for utilizing the available natural resource to improve the livelihood in a sustainable manner. Such Indigenous practices of different region needs some degree of social acceptance at the level of community. Gender plays an important role in accepting various practices. Peasants' own experiment with modern agricultural technology is often leading to better adoptability of the indigenous practices.

Such indigenous knowledge of land uses, pests, soil & water conservation, ethno-veterinary and ethno-medicine, etc. is similar to scientific knowledge. In many respects, it is better because it embodies years, if not generations, of specific practical knowledge which often also has the benefit of being interactive, holistic and not reductionist, thus incorporating the real linkages – consider crop testing under controlled conditions on an experimental farm, and in the actual farming system with real people (McCall, 2002). The indigenous practices, which are low-cost in nature and provide better economic advantage to the farmers, have comparatively better adoptability as against modern practices. Beyond this, there is indigenous knowledge that is apparently qualitatively different from scientific knowledge. This IK is symbolic, metaphoric, and visionary, (mystical in 'scientific' terms), and especially related with the land and land features.

ACKNOWLEDGEMENTS

This chapter is basically compilation information from various secondary sources for educational purposes. Therefore, thanks are due to all from whose sources the various valuable facts and figures are quoted.

References & Recommended Readings

Agrawal, A., 1995. Indigenous and scientific knowledge: some critical comments. *Indigenous Knowledge & Development Monitor* **3**: 3-5.

Ahmad, M., Khan, M. A. and Zafar, M., 2008. Traditional herbal cosmetics used by local women communities in district Attock of Northern Pakistan. *Indian Journal of Traditional Knowledge* **7**(3): 421-424.

Anonymous, 1977. Punjab Agricultural Hand Book, Punjab Agricultural University, Ludhiana, 111pp.

Arya, K. R. and Agarwal, S. C., 2008. Folk therapy for eczema, bone fracture, boils, sores and gingivitis in Taragtal province of Uttaranchal. *Indian Journal of Traditional Knowledge* **7**(3): 443-445.

Brouwer, R., 1995. The Measurement of the Non-Marketable Benefits of Agricultural Wildlife Management: The Case of Dutch Peat Meadow Land. Wageningen Economic Paper, 1995-1, Wageningen Agricultural University.

Campos, M. D. O., 2002. Etnociência ou etnografia de saberes, técnicas e práticas. In Métodos de Coleta e Análise de Dados em Etnobiologia, Etnoecologia e Disciplinas Correlatas. Amorozo, M. C. M., Ming, L. C., Silva, S. P., Rio, C. (eds.): UNESP/CNPq.

Choudhury, M., 1992. Fishing methods in flood plain lakes, In: Compendium of FAO sponsored training programme on flood plain fisheries management, CICFRI, Barrackpore, West Bengal, 163.

Das, A. K., Dutta, B. K. and Sharma, G. D., 2008. Medicinal plants used by different tribes of Cachar district, Assam. *Indian Journal of Traditional Knowledge* **7**(3): 446-454.

Deka, M. K., Bhuyan, M. and Harazika. K., 2006. Traditional pest management practices of Assam, *Indian journal of Traditional Knowledge* **5**(1): 75-78.

Dutta, R. and Bhattacharjya, B. K., 2009. A traditional fishing method of Assam for catfishes using duck meat as an attractant. *Indian Journal of Traditional Knowledge* **8**: 234-236.

El Ahry, M. T., 1988. Foreword Conservation Farming on Steep Lands. Soil and Water pp ix-xi in W.C. Moldenhauer Conservation Society of America, Ankey, I.A. and Hudson, N. (eds.).

Farnsworth, N. R., 1988. Screening of plants for new medicine. In: Biodiversity. Wilson, E. O. (ed) National Acad. Press. Washington. 63-97.

Howes, M. and Chambers, R., 1980. Indigenous knowledge: analysis, implications and issues. In: Brokensha, D. (ed.), Indigenous Knowledge and Development. University Press of America, Lanham, MD, pp. 323– 334.

Islam, M. N., Talukder, M. A.H., Rahman, M. L., Nasreen, A., Ali, S. M. M. and Banu, H., 2003. Comparative Efficacy of Some Indigenous Plant Materials as Toxicity and Repellent Against Pulse Beetle, *Callosobruchus chinensis* L. J. S. Pulido, G. Bocco/Geoderma, **111**: 249-265.

Jungerius, P.D., 1985. Perception and use of the physical environment in peasant societies. Geographical Papers, vol. 93. Department of Geography, University of Reading, England, 19pp.

Kalita, B., Dutta, A., Bhagabati, S. K. and Sharma, A., 2010. Indigenous Technical Knowledge for fish harvesting in Karbi-Anglong district of Assam. *Indian Journal of Traditional Knowledge* **9**(2): 252-255.

Kargioglu, M., Cenkci, S., Serteser, A., Evliyaoglu; N., Konuk, M., Kok, M. S. and Bagci, Y., 2008. An ethnobotanical survey of inner-West Anatolia, Turkey. *Hum. Ecol.* **36**: 763-777.

Lal, C. and Verma, L. R., 2008. Indigenous Technological Knowledge on soil and water management from Himachal Himalaya. *Indian Journal of Traditional Knowledge* **7**(3): 485-493

Lalthanzara, H. and Lalthanpuii, P. B., 2009. Traditional fishing methods in rivers and streams of Mizoram, north-east India. *Sci. Vis.* **9**(4):188-194.

McCall, M. K., 2002. Seeking Good Governance in Participatory- GIS. A review of Indigenous Spatial Knowledge, participatory processes, and governance dimensions in local- level GIS for development. Newer GISDECO.

McCall, M.K., 1995. Indigenous Technical Knowledge in Farming Systems of Eastern Africa: a Bibliography. Ames, IO: Iowa State University, Technology & Social Change Program, Bibliographies in Technology & Social Change No. 9. (167p.) (revised ed.).

McKinnon, J., 2001. Mobile interactive GIS: bringing indigenous knowledge and scientific information together. Paper at: Internet. Workshop on "Participatory Technology Development and Local Knowledge for Sustainable Land Use in Southeast Asia." Chiang Mai.

Murdoch, W. W., Chesson, J. and Chesson, P. L., 1985. Biological control in theory and practice, *Am. Nat.*, 125: 366.

Muthu, C., Ayyanar, M., Raja, N. and Ignacimuthu, S., 2006. Medicinal plants used by traditional healers in Kancheepuram district of Tamil Nadu, *India. J. Ethnobiol. Ethnomed.* **2**: 43.

Nasir Uddin, 2008. A study on the traditional housing technology of Bangladesh. *Indian Journal of Traditional Knowledge* **7**(3): 494-500

Nath, D. and Ray, D. C., 2012. Biological study of Red Pumpkin Beetle *Raphidopalpa foveicollis* (Lucas) on Cucurbita maxima during three crop seasons in Cachar district of Assam. Biodiversity: Researches in North East India. p. 115-127.

Oladele, O. I., Oladipo, O. A. and Ogunlade, I., 2008. Unintended consequences of arable crop technology within farming systems in Oyo state, Nigeria. *Indian Journal of Traditional Knowledgel* **7**(3): 429-435.

Omobuwajo, O. R., Alade, G. O. and Sowemimo, A., 2008. Indigenous Knowledge and practices of women herb sellers of Southwestern Nigeria. Indian Journal of Traditional Knowledge 7(3): 505-510.

Ozgen. U., Kaya, Y. and Coskun, M., 2004. Ethnobotanical studies in the villages of the district of Ilýca (Province Erzurum), Turkey. *Econ. Bot.* **58**:691-696.

Pathak, K.A., Thakur, N.S.A., Rao, K.R. and Shylesha, A.N., 2001. Insect pests of crops and their management, in: Steps Towards Modernization of Agriculture in NEH Region, ed N D Verma and B P Bhatt, (Indian Council of Agriculture Research, New Delhi), 121-159.

Posey, D., 1986. Introdução-Etnobiologia: Teoria e Prática. In Suma Etnológica Brasileira Edited by: Ribeiro B. Petrópolis: Vozes. 15-25.

Rahul, B. K., Mungole, A., Maske, M., Chaturvedi, A. and Chaturvedi, A., 2011. Indigenous traditional knowledge (ITK) from forest dwellers of Gondia district, *Maharashtra Journal of Ecobiotechnology* **3**(10): 14-18.

Rai, S. K. and Bhujel, R. B., 2002. Ethnic uses of some monocotyledonous plants in the Darjeeling Himalaya region, Proceeding Perspectives of Plant Biodiversity, Bishen Singh Mahendra Pal Singh, Dehradun.

Rajendran, K., Balaji, P. and Jothi Basu, M., 2008. Medicinal plants and their utilization by villagers in southern districts of Tamil Nadu. *Indian Journal of Traditional Knowledge* **7**(3): 417-420.

Richards, P., 1989. Farmers also experiment: a neglected intellectual resource in African science. *Discovery and Innovation* **1**(1):19-25.

Saha, R. K. and Nath, D., 2013. Indigenous Technical Knowledge (ITK) of fish farmers at Dhalai district of Tripura, NE India. *Indian Journal of Traditional Knowledge* **12** (1): 80-84.

Sen, S. K. and Behera, L. M., 2008. Ethnomedicinal plants used by the tribals of Bargarh district to cure diarrhoea and dysentery. *Indian Journal of Traditional Knowledge* **7**(3): 425-428.

Sharaby, A., 1988. Evaluation of some Myrtaceae plant leaves as protectants of rice against the infestation of *Sitophilus oryzae* L. and *Sitophilus granarius* L. *Insect Sci. Appl.* **9**: 465-468.

Simha, K. R. G., Laxminarayana, V., Prasad, S. V. L. N. and Khanum, S., 2008. Standardisation of *Yogaraja guggulu* — An Ayurvedic polyherbal formulation. *Indian Journal of Traditional Knowledge* **7**(3): 389-396.

Simsek. I., Aytekin, F., Yesilada, E. and Yildirimli, S., 2004. An ethnobotanical survey of the Beypazarý, Ayas, and Güdül district towns of Ankara province (Turkey). *Econ. Bot.* **58**: 705-720.

Singh, R. K. and Sureja, A. K., 2008. Management of indigenous sugarcane under rice-wheat based agroecosystem through informal institutions. *Indian Journal of Traditional Knowledge* **7** (3): 469-477.

Srivastava, S. K., Sarkar, U. K. and Patiyal, R. S., 2002. Fishing methods in streams of the Kumaon Himalayan region of India. *Asian Fish Sci.* **15**: 347-356.

wgbis.ces.iisc.ernet.in/energy/HC270799/LM/SUSLUP/.../76.pdf.

Wilken, G. C., 1987. Good Farmers: Traditional Agricultural Resource Management in Mexico and Central America University of California Press, Berkeley, CA.

Chapter – 2

Basic Concept of Indigenous Knowledge and it's Protection in India

Ajit Kumar Roy

INTRODUCTION

Knowledge is a familiarity with someone or something, which can include facts, information, descriptions, or skills acquired through experience or education. It can refer to the theoretical or practical understanding of a subject. It can be implicit (as with practical skill or expertise) or explicit (as with the theoretical understanding of a subject); it can be more or less formal or systematic. While many would agree that one of the most universal and significant tools for the transfer of knowledge is writing (of many kinds), argument over the usefulness of the written word exists nonetheless, with some scholars skeptical of its impact on societies. People are often internalizing new information which they perceive to be knowledge but in reality fill their minds with false knowledge.

Verbal communication lends itself to the spread of falsehoods much more so than written, as there is no record of exactly what was said or who originally said it (usually neither the source nor the content can be verified). Gossip and rumors are common examples. As to value of writing, the extent of human knowledge is now so great that it is only possible to record it and to communicate it through writing. Major libraries today can have millions of books of knowledge (in addition to works of fiction). It is only recently that audio and video technology for recording knowledge has become available and the use of these still requires replay equipment and electricity. Verbal teaching and handing down of knowledge is limited to those few who would have contact with the transmitter person - far too limited for today's world. Writing is still the most available and most universal of all forms of recording and transmitting knowledge. It stands unchallenged as mankind's primary technology of knowledge transfer down through the ages and to all cultures and languages of the world. Other biological domains where "knowledge" might be said to reside, include: (i) the immune system, and (ii) in the DNA of the genetic code. Such considerations seem to call for a separate definition of "knowledge" to cover the biological systems. For biologists, knowledge must be usefully available to the system, though that system need not be conscious. Thus the criteria seem to be:

- The system should apparently be dynamic and self-organizing (unlike a mere book on its own).
- The knowledge must constitute some sort of representation of "the outside world" or ways of dealing with it (directly or indirectly).
- Some way must exist for the system to access this information quickly enough for it to be useful (Anon., 2008).

Traditional knowledge (TK), indigenous knowledge (IK), traditional environmental knowledge (TEK) and local knowledge generally refer to knowledge systems embedded in the cultural traditions of regional, indigenous, or local communities. Traditional knowledge includes types of knowledge about traditional technologies of subsistence (e.g. tools and techniques for hunting or agriculture), midwifery, ethnobotany and ecological knowledge, celestial navigation, ethnoastronomy, etc. These kinds of knowledge are crucial for the subsistence and survival and are generally based on accumulations of empirical observation and interaction with the environment.

In many cases, traditional knowledge has been orally passed for generations from person to person. Some forms of traditional knowledge are expressed through stories, legends, folklore, rituals, songs, and even laws. Other forms of traditional knowledge are expressed through different means (Acharya and Shrivastava, 2008)

Traditional knowledge has been defined as "a cumulative body of knowledge, know-how, practices and representations maintained and developed by peoples with extended histories of interaction with the natural environment. These sophisticated sets of understandings, interpretations and meanings are part and parcel of a cultural complex that encompasses language, naming and classification systems, resource use practices, ritual, spirituality and worldview" (Anon., 2002).

DIFFERENT KINDS OF KNOWLEDGE

No matter what importance western knowledge system assumes in the industrialized countries, it is only one of the knowledge systems. It exists along with many non-formal indigenous (local, traditional or indigenous) knowledge systems in a lot of countries.

'Formal knowledge' refers to knowledge systems developed within the predominantly western-based education systems. It is called formal because it tends to be supported by written documents, rules and regulations, and technological infrastructure.

In contrast, the concept of "indigenous or local knowledge" (IK) refers to the knowledge, know-how, and practices developed and maintained by people with long histories of close interaction with their natural environment. (*Source*: http://www.scidev.net/dossiers/ index. cfm? fuseaction =dossierfulltext &Dossier=7)

Indigenous knowledge is fundamental to local decision-making regarding daily activities like hunting and gathering, fishing, agriculture, animal husbandry, water conservation, health, etc. Moreover, unlike formal scientific knowledge, indigenous knowledge is generally transferred as oral wisdom from one generation to the other, and is seldom, if ever, documented. Combining all forms of knowledge other than the formal ones as 'indigenous knowledge' would lead to its generalization and oversimplification, and may negate the outstanding contribution local knowledge can make to sustainable development.

VARIOUS DEFINITIONS OF INDIGENOUS KNOWLEDGE

Traditional knowledge (TK), also known as indigenous knowledge (IK) or local knowledge (LK) generally refers to the matured long-standing traditions and practices of certain regional, indigenous, or local communities. It also encompasses the wisdom, knowledge, and teachings of these communities. (Source: http://en.wikipedia.org/wiki/Indigenous_knowledge)

Indigenous knowledge (IK) has also been referred to as the unique, traditional, local knowledge existing within and developed around the specific conditions of women and men indigenous to a particular geographic area (Grenier, 1998).

IMPORTANCE OF INDIGENOUS KNOWLEDGE

The indigenous people of the world possess an immense knowledge of their environments, based on centuries of living close to nature. Living in and from the richness and variety of complex ecosystems, they have an understanding of the properties of plants and animals, the functioning of ecosystems and the techniques for using and managing them that is particular and often detailed. In rural communities in developing countries, locally occurring species are relied on for many - sometimes all - foods, medicines, fuel, building materials and other products. Equally, people's knowledge and perceptions of the environment, and their relationships with it, are often important elements of cultural identity.

Indigenous communities, peoples and nations are those which, having a historical continuity with pre-invasion and pre-colonial societies that developed on their territories, consider themselves distinct from other sectors of societies now prevailing in those territories, or parts of them. They form at present non-dominant sectors of society and are determined to preserve, develop and transmit to future generations their ancestral territories, and their ethnic identity, as the basis of their continued existence as peoples, in accordance with their own cultural patterns, social institutions and legal systems. Indigenous communities depend on their immediate environment to meet most of their basic needs. Therefore, they possess a deep appreciation of the environment and its underlying processes which forms the foundation for decision making in most day-to-day activities. (*Source*: Cited in *Global Issues* for the 90s, Centre for Teaching International Relations, University of Denver, 1993, pp. 86-97).

TRADITIONAL EDUCATION

Indigenous knowledge has been passed down from generation to generation through traditional education, with adults teaching practical knowledge of culture, the environment and survival through demonstrations and through a wide range of ceremonies, stories, songs, village meetings and taboos.

FORMAL EDUCATION

Formal education was introduced to many developing countries in the 19th century (often by colonial governments) to produce administrators, clerks, teachers and interpreters. This type of education was based on abstract knowledge systems – scientific knowledge – that evolved in the western industrialised world. Formal education systems had little place for indigenous knowledge or indigenous methods of education. It was, until recently, assumed that indigenous knowledge was irrelevant, unscientific and outdated. Therefore, few attempts were made to integrate indigenous knowledge into formal education despite its potential value in solving contemporary problems. As a result, education was confined to classrooms and children separated from their culture and environment. The teacher-centered nature of formal education also separated children from parents and, consequently, parents became less able to pass on the knowledge they had inherited to their children.

COMPARISON BETWEEN INDIGENOUS AND SCIENTIFIC KNOWLEDGE

Indigenous knowledge is different from non-indigenous knowledge, or 'scientific' knowledge, that most of us are familiar with. However, indigenous knowledge is still scientific. In the recent years, the role of IK in a range of sectors is being talked about. It includes intercropping techniques, pest control, crop diversity, and seed varieties in agriculture; plant varieties, and fish breeding techniques in biology; traditional medicine in human healthcare; soil conservation, irrigation, and water conservation in natural resource management; and oral traditions and local languages in education. The realization of IK's contribution to these sectors has led to an increasing interest in it by academicians, and policymakers alike.

Many government and non-governmental organizations, as well as international organizations such as the World Bank, International Labor Office, UNESCO and FAO are now appreciating the role IK can play in achieving sustainable development in a country. This interest is also apparent in the policies and programmes of various countries.

EROSION OF 'IK' SYSTEMS

Although some IK is lost naturally as practices get modified or are left unused for long time periods, the current rate of loss can be attributed to modernization and cultural homogenization, the current educational systems that believe macro-level problems can only be addressed through the global knowledge pool, and the slow growth of institution supporting grassroots innovations". *We are reminded of the global and historical tendency of complex technologies associated with economic powers to squash smaller, local technologies . . . We are urged to identify the valuable elements of smaller technologies and to create a place for them in the new century.*"(Kroma, 1996)

With rapid population growth, in-migration, and government relocation schemes (in case of large development projects), standards of living is often deteriorated. As poverty augments, short-term economical gains are chosen over environmental friendly local practices. The introduction of monocropping patterns in agriculture and forestry results in a loss in biodiversity, thus leading to a decline in IK. Deforestation leads to the disappearance of several precious yet unknown medicinal plants, and as a result the knowledge associated with those plants also declines. A lot of traditional knowledge is also being lost because of the communication gap, since neither children nor adults spend as much time in their communities. Because IK is generally transmitted orally, it is susceptible to change, particularly when people move to new regions, or when people's lifestyles tend to be different from that of their ancestors. Traditional knowledge typically distinguishes one community from another. For some communities, traditional knowledge takes on a personal and spiritual meaning. Traditional knowledge can also reflect a community's interests. Some communities depend on their traditional knowledge for survival.

Some social scientists conceptualize knowledge within a naturalistic framework, and emphasize the gradation of recent knowledge into knowledge acquired over many generations. These accounts use terms like "adaptively acquired knowledge", "socially constructed knowledge," and other terms that emphasize the social aspects of knowledge. Traditional knowledge, on the other hand, may be perceived very differently by indigenous and local communities themselves. The knowledge of indigenous and local communities is often embedded in a cosmology, and the distinction between "intangible" knowledge and physical things is often blurred. Indigenous peoples often say that "our knowledge is holistic, and cannot be separated from our lands and resources". Traditional knowledge in these cosmologies is inextricably bound to ancestors, and ancestral lands. Knowledge may not be acquired by naturalistic trial and error, but through direct revelation through conversations with "the creator", spirits, or ancestors (Chamberlin, 2003).

Indigenous and local communities often do not have strong traditions of ownership over knowledge that resembles the modern forms of private ownership. Many have clear traditions of custodianship over knowledge, and customary law may guide who may use different kinds of knowledge at particular times and places, and obligations that accompany the use of knowledge. From their perspective, misappropriation and misuse of knowledge may be offensive to traditions, and may have spiritual and physical repercussions in their cosmological systems.

Recently, international attention has turned to intellectual property laws to preserve, protect, and promote traditional knowledge. The reasons for this are complex. In 1992, the Convention on Biological Diversity (CBD) recognized the value of traditional knowledge in protecting species, ecosystems and landscapes, and incorporated language regulating access to it and its use (discussed below). It was soon urged that implementing these provisions would require revision of international intellectual property agreements.

This became even more pressing with the adoption of the World Trade Organization Agreement on Trade-Related Aspects of Intellectual Property Rights (TRIPs), which established rules for creating and protecting intellectual property that could be interpreted to conflict with the agreements made under the CBD. In response, the states who had ratified the CBD requested the World Intellectual Property Organization (WIPO) to investigate the relationship between intellectual property rights, biodiversity and traditional knowledge. WIPO began this work with a fact-finding mission in 1999. Considering the issues involved with biodiversity and the broader issues in TRIPs (involving all forms of cultural expressions, not just those associated with biodiversity - including traditional designs, music, songs, stories, etc.), WIPO established the Intergovernmental Committee on Intellectual Property and Genetic Resources, Traditional Knowledge and Folklore (IGC-GRTKF).

The period of the early 1990s to the Millennium was also characterized by the rapid rise in global civil society. The high-level Brundtland Report (1987) recommended a change in development policy that allowed for direct community participation and respected local rights and aspirations. Indigenous peoples and others had successfully petitioned the United Nations to establish a Working Group on Indigenous Populations that made two early surveys on treaty rights and land rights. These led to a greater public and governmental recognition of indigenous land and resource rights, and the need to address the issue of collective human rights, as distinct from the individual rights of existing human rights law.

The collective human rights of indigenous and local communities has been increasingly recognized - such as in the International Labor Organization (ILO) Convention 169 (1989) and the Declaration on the Rights of Indigenous Peoples (2007). The Rio Declaration (1992), endorsed by the presidents and ministers of the majority of the countries of the world, recognized indigenous and local communities as distinct groups with special concerns that should be addressed by states.

Initial concern was over the territorial rights and traditional resource rights of these communities. Indigenous peoples soon showed concern for the misappropriation and misuse of their "intangible" knowledge and cultural heritage. Indigenous peoples and local communities have resisted, among other things: the use of traditional symbols and designs as mascots, derivative arts and crafts; the use or modification of traditional songs; the patenting of traditional uses of medicinal plants; and the copyrighting and distribution of traditional stories.Indigenous peoples and local communities have sought to prevent the patenting of traditional knowledge and resources where they have not given express consent. They have sought for greater protection and control over traditional knowledge and resources. Certain communities have also sought to ensure that their traditional knowledge is used equitably - according to restrictions set by their traditions, or requiring benefit sharing for its use according to benefits which they define.Three broad approaches to protect traditional knowledge have been developed. The first emphasizes protecting traditional knowledge as a form of cultural heritage. The second looks at protection of traditional knowledge as a collective human right. The third, taken by the WTO and WIPO, investigates the use of existing or novel *sui generis* measures to protect traditional knowledge.

Currently, only a few nations offer explicit *sui generis* protection for traditional knowledge. However, a number of countries are still undecided as to whether law should give traditional knowledge deference. Indigenous peoples have shown ambivalence about the intellectual property approach. Some have been willing to investigate how existing intellectual property mechanisms (primarily: patents, copyrights, trademarks and trade secrets) can protect traditional knowledge. Others believe that an intellectual property approach may work, but will require more radical and novel forms of intellectual property law ("sui generis rights"). Others believe that the intellectual property system uses concepts and terms that are incompatible with traditional cultural concepts, and favors the commercialization of their traditions, which they generally resist. Many have argued that the form of protection should refer to collective human rights to protect their distinct identities, religions and cultural heritage.

Literary and artistic works based upon, derived from or inspired by traditional culture or folklore may incorporate new elements or expressions. Hence these works may be "new" works with a living and identifiable creator, or creators. Such contemporary works may include a new interpretation, arrangement, adaptation or collection of pre-existing cultural heritage that is in the public domain. Traditional culture or folklore may also be "repackaged" in digital formats, or restoration and colorization. Contemporary and tradition based expressions and works of traditional culture are generally protected under existing copyright law, a form of intellectual property law, as they are sufficiently original to be regarded as "new" upon publication. Once the intellectual property rights afforded to these new works of traditional knowledge expire, they fall into the public domain (Graber and Nenova, 2008). The public domain, as defined in the context of intellectual property rights, is not a concept recognised by indigenous peoples. As much of traditional knowledge has never been protected under intellectual property rights, they can not be said to have entered any public domain.

CONVENTION ON BIOLOGICAL DIVERSITY (CBD)

The Convention on Biological Diversity (CBD), signed at the United Nations Conference on Environment and Development (UNCED) in 1992, was the first international environmental convention to develop measures for the use and protection of traditional knowledge, related to the conservation and sustainable use of biodiversity. By 2006, 188 had ratified the Convention and agreed to be bound by its provisions, the largest number of nations to accede to any existing treaty (the United States is one of the few countries that has signed, but not ratified, the CBD). Significant provisions include:

Article 8. In-situ Conservation: Each Contracting Party shall, as far as possible and as appropriate: (a)... (j) Subject to its national legislation, respect, preserve and maintain knowledge, innovations and practices of indigenous and local communities embodying traditional lifestyles relevant for the conservation and sustainable use of biological diversity and promote their wider application with the approval and involvement of the holders of such knowledge, innovations and practices and encourage the equitable sharing of the benefits arising from the utilization of such knowledge, innovations and practices

Article 10. Sustainable Use of Components of Biological Diversity. Each Contracting Party shall, as far as possible and as appropriate protect and encourage customary use of biological resources in accordance with traditional cultural practices that are compatible with conservation or sustainable use requirements

At the Convention on Biological Diversity meeting, in Buenos Aires, in 1996, emphasis was put on local knowledge. Key players, such as local communities and indigenous peoples, should be recognized by States, and have their sovereignty recognised over the biodiversity of their territories, so that they can continue protecting it (Roussel and Synthèse, 2003)

The Protocol treats of inter-governmental obligations related to genetic resources, and includes measures related to the rights of indigenous and local communities to control access to and derive benefits from the use of genetic resources and associated traditional knowledge.

GOVERNMENT OF INDIA EFFORTS TO PROTECT TRADITIONAL KNOWLEDGE

In 2001, the Government of India set up the Traditional Knowledge Digital Library (TKDL) as repository of 1200 formulations of various systems of Indian medicine, such as Ayurveda, Unani and Siddha and 1500 Yoga postures (asanas), translated into five languages — English, German, French, Spanish and Japanese

(Anon., 2010a, 2010b) India has also signed agreements with the European Patent Office (EPO), United Kingdom Intellectual Property Office (UKIPO) and the United States Patent and Trademark Office (USPTO) to prevent the grant of invalid patents by giving patent examiners at International Patent Offices access to the TKDL database for patent search and examination (Anon., 2010c; 2010d).

'IPR' POLICY OF KERALA GOVT. (INDIA)

Intellectual Property Rights (IPRs) Policy for Kerala released in 2008, proposes adoption of the concepts 'knowledge commons' and 'commons licence' for the protection of traditional knowledge. The chief architects of the Policy are Prof. Prabhat Patnaik and Mr. R.S. Praveen Raj. The policy seeks to put all traditional knowledge into the realm of "knowledge commons", distinguishing this from the public domain. While codification of TK in digital libraries and sharing the same with patent offices prevents direct misappropriation, it is feared that it may provide an opportunity for private appropriation by making cosmetic improvements to such traditional knowledge that is not readily accessible otherwise. TKDL cannot at the same time be kept confidential and treated as prior art - says R.S. Praveen Raj

BEST PRACTICES ON INDIGENOUS KNOWLEDGE

'Best practices related to indigenous knowledge' refer to examples and cases that illustrate the good use of IK in developing cost-effective and sustainable survival strategies for poverty alleviation and income generation, e.g. indigenous land use systems to encourage labour-sharing arrangements among farmers; using IK to increase the fuel-efficiency of local stoves instead of replacing them; using indigenous institutions by extending credit through existing village loan groups etc. In collecting the information we are not interested in the details of the indigenous knowledge itself (the technical specifications of for example the stove are not relevant here), but in the ways that knowledge has been adapted, applied, and disseminated.

AIM OF THE DATABASE

The aim of this database is to encourage researchers and policy-makers to incorporate indigenous knowledge into their project proposals, feasibility studies, implementation plans and project assessments, and to take indigenous knowledge and practices into account in all activities affecting local communities. We know that many people are working on projects in which indigenous knowledge plays an essential and practical role. It is very important that information about this kind of projects is made available world-wide so that other people can learn from the experiences. The database of best practices will play an important role in building a bridge between empirical solutions, research and policy.

CHALLENGES OF MANAGING INDIGENOUS KNOWLEDGE RESOURCES IN INDIA

Introduction Indigenous knowledge is transferred from one generation to another through practice and to a large extent through oral communication and therefore remains confined to some areas, communities and individuals. Human beings used the natural resources like medicinal plants for treatment of disease and maintenance of health because of easy accessibility, reliability, efficacy, and affordability and least possible or no side effects of consumption of such materials. Besides herbal oils, breathing techniques, yogic asana, mantras, use of gemstones, several others indigenous systems exist for health and well being of human beings. Enormous numbers of medicinal plants based healers exist in the world and this knowledge is not generally widely accessible. Therefore whatever exists in the recorded form is just tip of the iceberg.

Indigenous Knowledge Resources in India

Indigenous medical knowledge in India is so diverse and complex that ranges from chanting of mantras, diagnosis by oracles, administration of holy water and biphutis for cure to use of sand of a specific place for cure of skin diseases. Indigenous medicinal knowledge is also practiced by different cultures for treatment prevention and cure of disease according to their cultural beliefs and folk medicine. Ayurveda system of medicine identifies the root cause of the disease and ensures holistic treatment of the human body instead of just curing the specific ailment. There are other systems such as Unani, Tibetan, Naturopathy and yoga for maintenance of health and well being of people besides the allopathy. There is a general belief that family of a home that plants neem (*Azadirchta indica*) never suffers from a diseases like tuberculosis. This plant is used in several ailments ranging from skin diseases to diabetes. Hindu families generally grow a plant of Tulsi (*Ocimum sativum*) at their homes both in rural and urban India and worship it. Tulsi is used in Ayurveda system of medicine for cure of several diseases. Beside the plant based medicinal knowledge some animal based systems and products have been developed in India. Olive green leeches had been traditionally used for purification of human blood as leech therapy in the unani system of medicine.

Publicity and preservation of indigenous knowledge

There is an enormous volume of indigenous knowledge existing in India. But it is cocooned in tribal communities and is widely dispersed and remains confined to local communities because of lack of publicity. Indigenous knowledge is also vanishing at an alarming rate. There is therefore an urgent need for its documentation, preservation of worthwhile knowledge, and its publicity.

Scientific validation of indigenous knowledge

Some Western researches have shown interest in scientific investigation and validation of indigenous knowledge resources especially for providing alternative medicine and supplementary therapies for cure of chronic diseases. Indigenous knowledge resources in a country like India are so diverse that in depth study of every case provides different and interesting information.

Reasons for disappearance of Indigenous Knowledge

Tacit domain, Multi-lingual forms, Wide spread illiteracy, Cultural barriers, Fears of bio-piracy, Idea robbery, Large scale urbanization, Break down of joint family system, Allurement of younger generation towards other lucrative work, Lack of recognition of practitioners of traditional knowledge, Extinction of plant species due to rapid deforestation and global warming, Fast life and changing life styles, Lack of awareness of patents and IPR among the custodians of indigenous knowledge.

Suggestions

Local people have wealth of information on usage of plants in their respective areas. They can provide adequate leads for scientific validation of their indigenous knowledge and develop suitable molecules for indigenous knowledge based drugs. For this more comprehensive databases of worthwhile indigenous knowledge are required to be created. Efforts are now necessary for capturing indigenous knowledge from some tribal and other communities by creating videotapes of especially illiterate practitioners of indigenous knowledge including show how kind of information. Such custodians of indigenous knowledge be incentavized and motivated to divulge their ideas in the larger interest of the humanity by protecting their IPRs. Linkages between universities and grassroots innovators, synergy between local communities, research institutions, pharmaceutical companies and other industries is essential for putting indigenous knowledge to greater use. There is a lot of room in a country like India for work on ancient wisdom from the perspective of modern science.

KNOWLEDGE MANAGEMENT PORTAL OF INDIA (www. rkmp. co. in– Directorate of Rice Research (www.drricar.org)

The new agricultural paradigm will have to be recast to take advantage of the wealth of knowledge available to achieve multiple goals of sustaining the food security, income, jobs, etc. In India, rice is the major cereal crop grown in about 44.6 million hectares. India represents all kinds of diversity under which rice is grown across the globe. No other crop is as versatile as rice. Given the burgeoning task of further enhancing the production and productivity of rice, the existing information sharing mechanisms appear to be insufficient. It is being increasingly felt that ICTs can be a major vehicle for all round socio-economic development. Given the task of reaching so many state department officials, NGOs, scientists working in rice regions of the country, it is not possible to cater the needs of all the stakeholders with the given resources. One of the viable options we thought of was developing and maintaining a Knowledge Management Portal on Rice that is expected to help millions of rice workers located in various parts of the country to be effective and efficient in collecting, organizing, summarizing, analyzing and synthesizing the rice knowledge helping in better decision making. In this changing scenario, Indian Council of Agricultural Research (ICAR) wishes to promote the use of Information and Communication Technologies (ICTs) in agriculture by developing national level knowledge management portals. As a first step towards achieving this objective, an exclusive portal on rice *viz.*, Rice Knowledge Management Portal (RKMP-www.rkmp.co.in) is developed by ICAR under the National Agricultural Innovation project (NAIP).

For providing the most comprehensive agricultural knowledge directly from the scientific community, 'Rice Knowledge Management Portal' is developed through a series of consultations involving the consortium partners and other ICT4D experts. In a short span of 24 months, the portal is built with high quality content developed by 8 consortium partners, 20 AICRIP centres and thousands of other stakeholders.

RKMP now serves as an information highway for sharing rice knowledge across the country. It has several global firsts in terms of comprehensiveness and utility. Built on web 2.0 standards, the portal helps any rice stakeholder to share their information for easily uploading the content into the portal. This helps in catering to location specific information needs of many stakeholders in local languages. With about 20 platforms, more than 10,000 pages of content, 3000 minutes of audio, 50 video clips, this is the first comprehensive agricultural portal of the country.

In research domain, various services are provided such as AICRIP Intranet, archives of 27000 datasets, bio-informatics suite, India Rice Research Repository (i3R) etc. It operates two e-learning platforms providing learning opportunity to scientists and extension workers at their time and space.

CONSERVATION OF TRADITIONAL KNOWLEDGE

International attention has in the past years has shifted to intellectual property laws to preserve, and promote traditional knowledge. The Convention on Biological Diversity of 1992 acknowledged the contribution traditional knowledge can make in protecting species, ecosystems and landscapes, and therefore included laws pertaining to its access and use. Indigenous people and local communities have since long resisted the use of art and crafts, traditional symbols and designs; the use or modification of traditional songs; the patenting of traditional uses of medicinal plants; and the copyrighting and distribution of traditional stories. Certain communities have also sought to ensure that their traditional knowledge is used equitably - according to restrictions set by their traditions, or requiring benefit sharing for its use.

Experience has proved that developmental activities that tend to ignore local technologies; local knowledge systems, and those which fail to formulate policies without a concern for local environment generally fail to achieve the desired outcomes.

NATIONAL INNOVATION FOUNDATION (NIF)

This is a unique initiative towards conserving indigenous knowledge systems and promoting grassroots innovations. NIF was launched in 2000 by the Government of India as India's national initiative to strengthen the Grassroots technological innovations and Traditional Knowledge. For the last seventeen years, the Honeybee Network and Society for Research and Initiatives for Sustainable Technologies and Institutions (SRISTI) have been scouting innovations by farmers, artisans, women, etc. at the grassroots level. The Gujarat Grassroots Innovations Augmentation Network (GIAN) scales up innovations, from the Honey Bee database of innovations, through value additions in innovations to sustain creativity and ethics of experimentation. Following are the steps taken by these organizations in order to ensure a unique platform for grassroots innovations:

- Scouting and documentation of innovations: This requires extensive fieldwork, travel in rural and urban areas, search for 'odd balls' — the experimenters – and local community and knowledge experts in the society
- Knowledge dissemination: The entire effort of scouting and documentation leads to the development of a database of innovations and traditional knowledge aimed at building the electronic National Register. The process involves using various information technology and database applications for horizontal networking amongst innovators and traditional knowledge experts as well as other stakeholders.

- Value addition and R&D: Most of the innovators and/or traditional knowledge experts need optimization in design and/or product formulation through blending with modern science and technology inputs. Market prospects for many innovations will be very low without proper value addition. Efficiency gains can be made by creating technology networks.
- Business development and micro venture.
- Learning: Cross regional, peer-to-peer learning exchanges aimed at enhancing local capacity to identify and apply indigenous/traditional knowledge and practices.

(*Source*: http://nifindia.org/activity.htm)

INDIGENOUS INTELLECTUAL PROPERTY

Indigenous intellectual property is an umbrella legal term used in national and international forums to identify indigenous peoples' special rights to claim (from within their own laws) all that their indigenous groups know now, have known, or will know(Anon., 2004). It is a concept that has developed out of a predominantly western legal tradition, and has most recently been promoted by the World Intellectual Property Organisation, as part of a more general United Nations push to see the diverse wealth of this world's indigenous, intangible cultural heritage better valued and better protected against probable, ongoing misappropriation and misuse.

In the lead up to and during the United Nations International Year for the World's Indigenous Peoples (1993) then during the following United Nations Decade of the World's Indigenous Peoples (1995–2004) a number of conferences of both indigenous and non-indigenous specialists were held in different parts of the world, resulting in a number of declarations and statements identifying, explaining, refining, and defining 'indigenous intellectual property'.

AGREEMENT ON TRADE-RELATED ASPECTS OF INTELLECTUAL PROPERTY RIGHTS (TRIPS)

Agreement on Trade-Related Aspects of Intellectual Property Rights (TRIPs) sets out certain conditions under which certain biological materials or intellectual innovations may be excluded from patenting. The Article also contains a requirement that Article 27 be reviewed. In the TRIPs-related Doha Declaration of 2001, Paragraph 19 expanded the review to a review of Article 27 and the rest of the TRIPs agreement to include the relationship between the TRIPS Agreement and the 1992 Convention on Biological Diversity (CBD) and the protection of traditional knowledge and folklore.

ACKNOWLEDGEMENTS

This chapter is basically compilation information from various secondary sources for educational purposes. Therefore, thanks are due to all from whose sources the various valuable facts and figures are quoted.

References and Recommended Readings

Acharya, D. and Shrivastava, A., 2008. Indigenous Herbal Medicines: Tribal Formulations and Traditional Herbal Practices, Aavishkar Publishers Distributor, Jaipur- India. ISBN 978-81-7910-252-7. pp 440.

Anon., 1993. Rainforest Aboriginal Network (1993) Julayinbul: Aboriginal Intellectual and Cultural Property Definitions, Ownership and Strategies for Protection. Rainforest Aboriginal Network. Cairns. Page 65.

Anon., 2002. International Council for Science / Conseil International pour la Science (March 2002). "Science and Traditional Knowledge: Report from the ICSU Study Group on Science and Traditional Knowledge". p. 3. http://www.icsu.org/publications/reports-and-reviews/science-traditional-knowledge/Science-traditional-knowledge.pdf. Retrieved 24 May 2012.

Anon., 2004.World Bank. *Indigenous knowledge: local pathways to global development*, Africa Regional Office, World Bank.

Anon., 2007. Office of the United Nations High Commissioner for Human Rights (2007). "Indigenous peoples, Geneva. Archived from the original on 2007-11-07. http://web.archive.org/web/20071107144031/http://www.ohchr.org/english/issues/indigenous.

Anon., 2008. IPRs policy proposes 'knowledge commons'. The Hindu (Chennai, India). 28 June 2008. http://www.hindu.com/2008/06/28/stories/2008062856600100.htm.

Anon., 2010a. Know Instances of Patenting on the UES of Medicinal Plants in India. PIB, Ministry of Environment and Forests. May 6, 2010.

Anon., 2010b. CSIR wing objects to Avesthagen patent claim. Live Mint. April 28, 2010. http://www.livemint.com/2010/04/28214947/CSIR-wing-objects-to-Avesthage. html.

Anon., 2010c. India Partners with US and UK to Protect Its Traditional Knowledge and Prevent Bio-Piracy". Press Information Bureau, Ministry of Health and Family Welfare. April 28, 2010. http://pib.nic.in/release/release.asp?relid=61122. Retrieved 25 May 2010.

Anon., 2010d. Know Instances of Patenting on the UES of Medicinal Plants in India". PIB, Ministry of Environment and Forests. May 6, 2010.

Bates, P., Chiba, M., Kube, S. and Nakashima, D., 2009. Learning and Knowing in Indigenous Societies Today, UNESCO, Paris.

Chamberlin, J. E., 2003. If This Is Your Land, Where Are Your Stories? Finding Common Ground. Toronto: Alfred A. Knopf Canada.

Conway, K., 1997. Improving crop resistance: a new plant breeding technique borrows from the past. IDRC Reports, 2 May 1997. http://archive.idrc.ca/books/reports/1997/17-01e.html.

Graber, C. B. and Nenova, M. B., 2008. Intellectual Property and Traditional Cultural Expressions in a digital environment. Edward Elgar Publishing. pp. 174.

Grenier, L., 1998. *Working with Indigenous Knowledge: A Guide for Researchers*, IDRC, Canada. http://www.idrc.ca/en/ev-28703-201-1-DO_TOPIC.html.

http://en.wikipedia.org/wiki/Indigenous_knowledge.

http://pib.nic.in/release/release.asp?relid=61511.

http://www.nufficcs.nl/ciran/ikdm/.

http://www.scidev.net/dossiers/index.cfm?fuseaction=dossierfulltext&Dossier=7.

IFAD, 2003. *Indigenous Peoples and Sustainable Development*, Roundtable Discussion Paper for the Twenty-Fifth Anniversary Session of IFAD's Governing Council.

Johnston, A. M., 2005. *Is the Sacred for Sale. Tourism and Indigenous Peoples*, Earthscan, London.

Kroma, S., 1996. The science of Pacific Island peoples. Indigenous Knowledge and Development Monitor 4(2). http://www.nufficcs.nl/ciran/ikdm/.

Le Grange, L., 2007. 'Integrating Western and Indigenous Knowledge Systems: The Basis for Effective Science Education in South Africa?', *International Review of Education*, **53**(5-6): 577-591.

Peña, D. G., 2005. Mexican Americans and the Environment, The University of Arizona Press, p. 198.

Posey, D. A., 2008. Indigenous Management of Tropical Forest Ecosystems: The case of the Kayapo Indians of the Brazilian Amazon" in Dove and Carpenter, Environmental Anthropology: a historical reader, Blackwell Publishing, Oxford, p. 90.

Roussel, B. and Synthèse, 2003. The CBD: Local knowledge systems at the heart of international debates, Institute for Sustainable Development and International Relations.

Sillitoe, P., 2009. *Local Science Vs Global Science: Approaches to Indigenous Knowledge in International Development*, Berghahn Books.

Wasongo, V. O., Kambewa, D., and Bekalo, I., 2011. Community-Based Natural Resource Management (p. 194). In W.O. Ochola, P.C. Sanginga, I. Bekalo (Eds.), Managing Natural Resources for Development in Africa. A Resource Book. Nairobi, Kenya: Univ. of Nairobi Press, 2011 (Chapter 4: pp. 165-210).

Watson, Irene, 1993. International Year for Indigenous Peoples. Aboriginal Law Bulletin. AustLII. http://www.austlii.edu.au/au/journals/AboriginalLB/1992/52.html.

www.gian.org.

www.nifindia.org.

www.sristi.org.

Zweifel, H., 1997. Biodiversity and the appropriation of women's knowledge. *Indigenous Knowledge and Development Monitor* **5**(1).

Chapter – 3

Traditional Ecological Knowledge: An Important Pillar in Fisheries Resource Management

Biswarup Saha, Tanmay Samajdar and P. K. Pal

INTRODUCTION

Sustainable exploitation of fish stocks and conservation of fish habitat are matters of concern for fishery scientists and administrators all over the world. Worldwide there is a declining trend in natural fish catch (FAO, 2002) and India is no exception. Increasing harvest pressure, destructive fishing operations, introduction of exotics, are examples of factors from within the sector responsible for increasing pressure on fish stocks. The decline in open water catch has been a serious issue in fisheries as millions of rural households depend on fishing to maintain their livelihood incomes, and because fish is a major protein source in the globe. However, the government in several countries is addressing the issue in a number of ways, such as: (i) moving towards biological management from revenue-oriented management; (ii) involving resource users in capture fisheries management to maintain natural productivity; and (iii) starting programs such as rehabilitating of degraded habitats (FAO, 1999).

Fishing takes place according to habitat features, seasonality, movement, patterns, size and abundance of fish. Fishers engaged themselves with fisheries activities more times than the fisheries managers and scientists do. It is widely believed that modern fisheries knowledge is inadequate for solving present complex fisheries problems. Given the long term as well as historical livelihood relevance of fishers with water body dynamics and fishing, it is natural that fishers possess a vast knowledge with regard to stocks and habitats (Neis et al., 1996). Understanding fisheries resource dynamics with respect to biophysical characters and identifying management problems with the help of the traditional fisher knowledge might help sustainable management. Fishers' knowledge also can help qualitative and quantitative characterization of the fisheries. Incorporating local fisheries knowledge into conventional management with respect to ecology and biology may ensure a thorough approach to management. It will benefit both the communities and managers dealing with issues of habitat conservation and restoration. The incorporation of fishers' ecological knowledge into conventional management systems has the potential to help fisheries researchers and managers to understand the ecology and fisheries resources and can help build credibility with local communities. Building credibility could help in creating shared vision in fisheries management. The significance of local ecological knowledge in fisheries management has been studied in details in countries such as Norway, Canada, Newfoundland, Bangladesh and various Pacific island states. Exploring local knowledge among Sami fjord fishermen in northern Norway, Eythorsson (1993, 1998) revealed that local traditional knowledge and local norms could prove to be an important, or even a necessary, supplement to scientific knowledge for the establishment of ecologically sound and socially justified management of common property resources. Research has also highlighted the importance of local ecological knowledge in promoting community participation in fisheries management (Neis, 1992, McCay and Finlayson, 1995, McGoodwin et al., 2000). Few studies emphasized on the richness of fisher ecological knowledge in Bangladesh and highlighted the importance of incorporation of fisher knowledge into mainstream management and argued that it may play an important role in restoring degraded floodplain fisheries of Bangladesh (Sillitoe, 2000, Thomson, 2003).

Traditional ecological knowledge (TEK) is the knowledge used by the local users to exploit ecosystem goods and services and also to respond to the change in the ecosystem process. In this sense, such TEK is also local knowledge and the terms are often used interchangeably. However, Berkes (1999) contends local knowledge is distinguished from TEK by virtue of the fact it has a short history—that is, it lacks tradition. He further emphasized that local knowledge does not have multigenerational records. But certainly local knowledge of today may be used as TEK at some point if it is continuously used by local people to meet their livelihood needs. TEK evolved as adaptive processes handed down through generations by cultural transmission. It is concerned with the relationship of living beings (including humans) to one another and to their environment. It is also noted that traditional ecological knowledge is both cumulative and dynamic, building on experience and adapting to changes (Johnson, 1992). It deals with local ways of knowing and being holistic in nature, helps to eliminate some of the shortcomings found in the continual development approach.

Traditional knowledge is often holistic in nature and based on personal observation following trial and error, an approach which is capable of dealing with complex social and ecological issues over the long term (Berkes, 1999). As a whole TEK is a multidisciplinary approach that is learned through observation and hands-on experience in an adaptive learning process and is more effective in dealing with uncertainly and complexity in management systems (Berkes, 1999, Berkes and Folke, 2002, Gadgil et al., 2003). Increasingly, TEK or local ecological knowledge (LEK) held by groups of people engaging directly with their ecosystems is recognized as a valuable tool for understanding environmental change, as well as for ecosystem management and conservation (Gracta-Quijano, 2007). Local knowledge, in fisheries context, means the understanding the fishers have about ecosystem processes including their awareness and skills gained through experience and day-to-day interactions with the environment. Vertical and horizontal transmission of fishing knowledge from one fisher to another is very common in local communities for a number of reasons such as cultural connections and local rule systems concerning use of a particular resource. However, the transmission of fisheries knowledge to scientists/managers is not always apparent and therefore, a formal arrangement for integration of these two spheres of knowledge is needed before using TEK in formal environmental and resource management.

NATURE OF TRADITIONAL ECOLOGICAL KNOWLEDGE

Many studies revealed that so far there is no universally accepted definition of traditional ecological knowledge (Grenier, 1998, Berkes, 1999) and the complexities of defining traditional ecological knowledge lie in the ambiguity of the words "traditional" and "ecological".

Berkes (1999) in his pioneering book Sacred Ecology, outlined the terms regarding traditional knowledge as follows:

1. **Traditional Ecological Knowledge:** A cumulative body of knowledge, practice, and belief, evolving by adaptive processes; it is handed down through generations by cultural transmission. It is about the relationship of living beings (including humans) with one another and with their environment. It is also mentioned that traditional ecological knowledge is both cumulative and dynamic, building on experience and adapting to changes. It deals with local ways of knowing and holistic in nature, thus helping to

eliminate shortcomings of the continual development approaches and it provides supports towards political and economic gain of the local people.

2. **Local Knowledge**: Recent practical knowledge with short history can simply be called local knowledge. Such knowledge does not have multi-generational records.

3. **Indigenous Knowledge:** Traditional or local knowledge held by an indigenous group which is unique to a given culture or society. According to Grenier (1998) indigenous knowledge refers to the unique, traditional and local knowledge existing within, and developed around, the specific condition of women and men who are indigenous to a particular geographic area. Indigenous knowledge is parochial, confined to a small area, limited to what people can sense, observe and comprehend, using their own terms and concepts. Indigenous knowledge also referred to as "traditional" or "local" knowledge, as it is embedded in the community and is unique to a given culture, location or society (Ibid). Still, there lies confusion in the minds of many about the use of two words, "indigenous knowledge" and "traditional knowledge". Indigenous knowledge is that knowledge that is held and used by a people who identify themselves as indigenous of a place based on a "combination of cultural distinctiveness and prior territorial occupancy relative to a more recently-arrived population with its own distinct and subsequently dominant culture". Traditional knowledge is, on the other hand, that which is held by members of a distinct culture and/or sometimes acquired "by means of inquiry peculiar to that culture, and concerning the culture itself or the local environment in which it exists." UNEP/CBD/TKBD (1997). Traditional peoples as those who hold an unwritten corpus of long-standing customs, beliefs, rituals and practices that have been handed down from previous generations. They do not necessarily have claim of prior territorial occupancy to the current habitat; that is, they could be recent immigrants. Thus, Parasar (1994) stated rightly that "all indigenous knowledge is traditional by nature but not the vice-versa is true". Indigenous knowledge fits neatly in the traditional knowledge category but traditional knowledge is not necessarily indigenous (Fig.1).

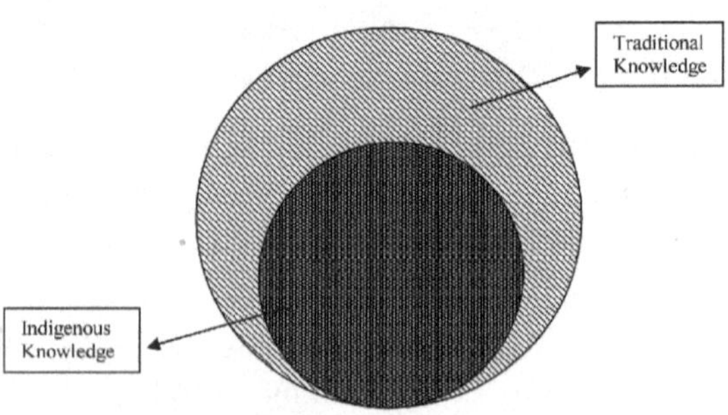

Fig. 1: Traditional knowledge system

The logic of the statement is that a knowledge developed locally may have scope to be transmitted down from generation to generation which after some years may be given the status of tradition, or considered as traditional knowledge. On the other hand, traditional knowledge, not necessarily have to possess an indigenous base. There is also the probability that it may have an exotic base which may have been introduced in the locality several years earlier through acculturation. While, in this way, some traditional knowledge has an exclusive indigenous base, some yet have both the indigenous and exotic components in them. Although the difference between the two terms is too thin and indiscernible to sustain in all conceivable cases, there may not be any semantic confusion in using the two expressions, 'traditional knowledge' and 'indigenous knowledge', somewhat interchangeable (Parasar, 1994).

IMPORTANCE OF TRADITIONAL ECOLOGICAL KNOWLEDGE

In recent years, there has been a growing awareness that the conventional approach to managing fisheries is insufficient. It is being recognized that sustainable use of the world's living aquatic resources can only be achieved if both the impacts of the ecosystem on the living resources and the impacts of the fishing on the ecosystem are explicitly identified and, as far as possible, understood (FAO, 2002). It is also being formally acknowledged that fishers are an integral part of the ecosystem and that their knowledge must be acknowledged to achieve wellbeing of both ecosystem and human (Neis et al., 1996).

In the socio-economic context, fishers' knowledge is seen in all steps of fish production, harvesting, handling or processing, marketing and consumption. Similarly, in the ecological context, fishers have diversified knowledge about fisheries resources such as fish habitat, bottom features, water current, nature and scales of fish aggregation and abundance and conservation of biodiversity (Bergmann et al., 2004), location of breeding ground and migration (Poulsen, 2001) and feeding and breeding biology of riverine fish (Faulkner et al., 2001). Local ecological knowledge helps the fishers identify the most productive fishing spots and are instrumental in determining the limits of the fishing area; maintaining fisheries yield and providing the basis for institutional rules regulating the spatial and temporal limitations for each users and helping with the conservation of fish (ibid).

Information possessed by fishers can be used to monitor stocks and physical environments, provide biological and technical data, and provide contextual information for management purposes. Fishers may have the knowledge on how the stocks is reduced More importantly, fishers are aware of contextual information useful in interpreting landings, biological and catch per unit effort data, information useful in understanding and assessing fisheries data (Grant and Berkes, 2004).

SCOPE OF TRADITIONAL ECOLOGICAL KNOWLEDGE

Fishers and other community members around the water body possess adequate knowledge with respect to fish habitats that could be used to further define the habitats. It was observed that fishers and local people define the habitats mainly on the basis of water depth, colours, water movements, location of the water body and its contribution to fish catch and habitat supports for fish. Therefore, TEK can be used to classify the fish habitats.

In addition to knowledge regarding fish habitat preference, fishers have adequate knowledge with respect to fish movement/migration between habitats. The information encompasses fish migration, growth and spawning techniques and biodiversity of fishes can be studied through TEK.

The hydrological cycle is also important for fish movement among habitats. The water availability and its impact on water quality and different techniques of fishing according to the traditional knowledge of fish habitats may be helpful through documentation of TEK. Over the years, habitat degradation, with respect to natural fisheries in India, has become an important policy concern, but a true solution has not been achieved yet. In the scientific community, as well as in management levels, there is enough technological knowledge and measures for the problems with respect to degradation. The study of fishers TEK may be found to be convergent about reasons for the degradation of fish habitats and its impact on fish production, livelihoods and biodiversity.

INDIGENOUS KNOWLEDGE VERSUS KNOWLEDGE

There are wider debates on indigenous knowledge versus western science-based techno-ecological knowledge. The orientation to indigenous knowledge systems is in stark contrast to the views of many earlier theorists, who saw traditional knowledge and institutions as obstacles to development and invited our attention that the human knowledge systems may be classified into two kinds: formal scientific knowledge (SK) system and traditional knowledge (TK) system. The main difference between these two kinds of knowledge systems is in their format. The SK system is essentially in explicit format – can be articulated in formal language including grammatical statements, mathematical expressions, specifications, manuals, and so forth. This kind of knowledge thus can be transmitted across individuals formally and easily. This has been the dominant mode of knowledge according to the (Western) scientific philosophy. However, the format of TK system is mostly tacit – hard to articulate with formal language. This knowledge is embedded in the experiences of indigenous or local people and involves intangible factors, including their beliefs, perspectives, and value systems.

Knowledge and understanding of Indigenous Knowledge System (IKS) combined with knowledge and understanding of Western agriculture systems can result in an emic-etic approach to agricultural education. An emic analysis of behaviour or phenomenon is based on internal structural or functional elements of a particular cultural system. An etic analysis is based on predetermined general concepts external to that cultural system.

An emic-etic perspective includes the recognition of other epistemologies, and of the holistic and integrated nature of IKS. A curriculum for training agricultural and extension educators to operate in the emic-etic perspective provides an ideological basis for learning with and from the people. The following highlights the special features of indigenous knowledge, which distinguishes it broadly from other knowledge. According to the literature (Ellen and Harris, 1996), IK is:

Local, in that it is rooted in a particular community and situated within broader cultural traditions; it is a set of experiences generated by people living in those communities. Separating the technical from the non-technical, the rational from the non-rational could be problematic. Therefore, when transferred to other places, there is a potential risk of dislocating IK.

Tacit knowledge and, therefore, not easily codifiable.

Transmitted orally, or through imitation and demonstration. Codifying it may lead to the loss of some of its properties.

Experiential rather than theoretical knowledge. Experience and trial and error, tested in the rigorous laboratory of survival of local communities constantly reinforce IK.

Learned through repetition, which is a defining characteristic of tradition even when new knowledge is added. Repetition aids in the retention and reinforcement of IK.

Constantly changing, being produced as well as reproduced, discovered as well as lost; though it is often perceived by external observers as being somewhat static.

Table 1: Some of the common differences between indigenous knowledge and western-based scientific knowledge

Sl.No.	Character	Local indigenous knowledge	Western scientific knowledge
1.	Data creation	Based on personal observations, trial and error, and synthesis of facts	Based on experimentation and systematic, deliberate accumulation of facts
2.	Data type	Qualitative	Quantitative
3.	Data generation	User involved	Specialized cadre of researchers
4.	Dominant mode of thinking	Intuitive, holisticMind and matter considered together	Analytical, reductionistMind reduced to matter
5.	Instruction	Learned through observation or hands on experience (Adaptive learning)	Got taught and learned in a situation usually separated from the applied context. Adaptive learning mostly absent.
6.	Effectiveness	Slow, inconclusive	Fast, conclusive
7.	Situation	Local context	Institution based
8.	Transmission	Oral or cultural means of communication and multi-generational	Limited to one-generation through One way using powerful media

BLENDING INDIGENOUS KNOWLEDGE SYSTEMS WITH OTHER FORMS OF KNOWLEDGE

According to Dean (1999) the main goal of technology blending is to both maintain local knowledge and supplement it; "expanding it and combining it with modern knowledge and technologies". He goes on to argue that it provides self sustaining livelihood for local people.

Local knowledge and technology must be amenable to commercial orientation at least to some degree in order to serve as a means to secure a livelihood. It will lead to more equitable sharing of the resulting benefits and increase the standard of living in traditional societies. This will be as a consequence of up-grading traditional production methods, elevating productivity, income and quality of life in the traditional sector. World Bank (1998) forwarded the following steps to integrate IK with other forms of knowledge:

1. IK has to be recognized, identified and selected from a multitude of other knowledge.
2. IK's validation/affirmation by identifying its significance, relevance, reliability, functionality, effectiveness and transferability. This signifies an ability to support problem solving.
3. Codification/recording/documentation.

4. Storage of IK for retrieval.
5. IK transfer. Such transfers go beyond focusing on human recipients.
6. Dissemination and use of IK. Use of knowledge put it to testing, acceptance and further validation for development.

Rajeshekaran (1993) has advocated a framework for incorporating IK into agricultural research and extension with the following features.

1. Strengthening the capacities of regional research and extension organizations;
2. Building upon local people's knowledge that are acquired through various processes such as farmer to farmer communication and farmer experimentation;
3. Identifying the need for extension scientist/social scientist in an interdisciplinary regional research team;
4. Formation of a sustainable technology development consortium to bring farmers, researchers, NGOs and extension workers together well ahead of the process of technology development;
5. Working with the existing organization and management of research and public sector extension;
6. Bringing research-extension-farmer together at all stages is practically difficult considering the existing bureaucracies and spatial as well as academic distances among the personnel belonging to these organizations. Hence the need for utilizing the academic knowledge gamed by some extension personnel (Subject Matter Specialists) during the process of validating farmer experiment.
7. Outlying areas that research and extension organizations need to concentrate on during the process of working with farmers;
8. Understanding that it is impractical to depend entirely on research stations for innovations; considering the inadequate human resource capacity of the regional research system.

Parasar (1994) developed a working model for incorporating indigenous technological knowledge in agricultural research and extension work:

1. Establishing a centre for indigenous knowledge
2. Giving curriculum sanction for indigenous knowledge Having a broad based extension approach

3. Framing appropriate policy
4. Seeking support from general academia
5. Involving NGOs
6. Giving wide publicity

COLLECTION, DOCUMENTATION AND VALIDATION OF TRADITIONAL ECOLOGICAL KNOWLEDGE SYSTEMS

In the second green revolution era, the concept of natural farming has gained importance. It is basically the development of the farmers themselves by using their own knowledge base and resources as well as their own innovativeness. It has been emphasized that a comprehensive understanding of the traditional knowledge and expertise is essential for sustainable development. To gain functional understanding of TEK system includes collection and assessment (validation) of such TEKs and finally its documentation.

Collection of TEK

Although in many discussions primitive societies are stressed as a source of IK but every society have its own indigenous knowledge systems which transmit from generation to generation. In many societies, although, IKs cannot be observed as an independent identity but these are interwoven with many of its package of practices. An investigator, for collection of IKs, has to employ an appropriate method of data collection/data gathering and have to identify appropriate sources of it. Pal (1998) had employed 'snow-ball' sampling to identify the sources of IKs in a society. He suggested that to employ snow-ball sampling technique, selection of the very first respondent is a very crucial one. However, the following characteristics may be considered for the selection of such respondents:

1. Good knowledge about the historical background of food production and resource conservation of the study villages,
2. A minimum of ten years of farming experience, and
3. Not being involved in other stages of the study.

The Mission Mode Project on "Collection, Documentation and Validation of Indigenous Technical Knowledge (ICAR)" has proposed the following sources of IK systems:

Table 2: Sources of Traditional Ecological Knowledge

	Primary sources		Secondary sources
1.	Fishers	1.	Extension workers and others
2.	Elderly persons of the area /village	2.	Ancient literature
3.	Village Mukhia	3.	Media (print and electronic)
4.	Retired persons	4.	Folksongs, proverbs & festivals
5.	NGO's/ Institutions		

Pal (1998) showed that among different sources of communication of ITK, forefathers are the strongest of all. Except forefathers, villagers communicated with progressive farmers and friends/relatives/neighbours channel for getting information regarding a traditional knowledge.

Different authors have suggested following methods for collection of ITK (Colfer, 1990, Mane and Sutaria, 1993, Quiroz, 1993, Rajesekaran, 1993, Pal and Das Gupta, 1997)

- Observation and interviewing
- Case studies
- Field observation
- Participant observation
- Interpretative research (IR)
- Local taxonomy
- Cognitive method analysis
- Crop history
- Innovator workshop
- Different PRA methods
- Participative technology analysis
- Illustrations and diagrams
- Mapping/area domains of the ITK
- Audio-visual/photo's/slides
- Use of schedule/questionnaire etc.

Training project report on Methodology of Collection, Documentation and Validation of ITKs (ICAR) (July, 17 to Aug, 6, 2004, CRIDA, Hyderabad, India) have presented the following methods also except the mentioned above. Those are:

VOLUNTARY DISCLOSURES/ REPORTING METHOD

Keeping in view the magnitude of the ITK documentation and paucity of existing database on the subject, one approach may be to invite the farmers/agencies to provide information in their respective area domains in various aspects of agriculture and crop husbandry. For this, appropriate formats require to be developed and circulated widely through advertisements using different media.

Visual aided dialogue

'Visual-aided dialogue' is another good method for researchers with farmers and for farmers with researchers. This involves discussing items, which are in front of the participants. Insect pests can be shown to farmers to elicit comments on pest control, or farmers can display and discuss the varieties of seeds in their stores and discuss their preferences and reasons for them. Using such visual aids has been found useful at the beginning of a research process, to break the ice and establish rapport, to stimulate interest, curiosity and discussion and to enable researchers to learn from farmers.

Manual discriminate analysis

Another method is manual discriminate analysis. This can be used in group discussions among farmers and researchers about farming practices and constraints. The basic concept is to compare and contrast. Farmers are asked to describe their management practices, for example for tillage, cropping patterns, input use, or techniques for maintaining soil fertility. Data are collected to find out the range of plot wise practices. The researcher then contacts the farmers whose practices are at the two contrasted ends of the distribution. Each group is asked to explain the behaviour of the other. For instance, farmers using the highest seed rates are asked about farmers with the lowest seed rates, and vice versa. This calibrates the frame of reference and makes clearer to farmers the distinctiveness of their own behaviour. Only after hypothesizing why the others behave differently in the group asked to explain the reasons for its own practices.

Methodology for documentation of TEK

In the WTO era when IPR related inter-country and inter-organization debates are emerging out, documentation of own knowledge systems, innovativeness or practices may be the solution for the same. In the domain of IK studies documentation may be done by categoring the collected IKs into different classes. NATP Mission Mode Project on "Collection, Documentation and Validation of Indigenous Technical Knowledge (ICAR)" has identified 24 thematic areas for documentation of ITK. Those are:

Table 3: Theme areas for documentation of Indigenous Knowledge

Sl. No.	Theme	Sl, No.	Theme
1.	Rainwater management	13.	Fisheries
2.	Soil and water conservation	14.	Ethno-botany and agro-bio-diversity
3.	Wind erosion	15.	Methods of weather forecasting
4.	Tillage practices	16.	Fuel management
5.	Crop and cropping systems	17.	Thermal efficiency
6.	Pest and disease management	18.	Waste water management
7.	Soil fertility management	19.	Garbage disposal and management
8.	Farm implements	20.	Food product development
9.	Post-harvest technology	21.	Natural yarns, dyes and weave
10.	Grain/ seed storage	22.	Low- cost housing material
11.	Horticultural crops	23.	Ethnic food
12.	Veterinary and animal husbandry	24.	Unclassified

Note: This list should not be considered as exhaustive

Documentation may be done with the help of the following methods as proposed in the training project report on Methodology of Collection, Documentation and Validation of ITKs (ICAR) (July, 17 to Aug, 6, 2004, CRIDA, Hyderabad, India):

1. Audio
2. Visual
3. Audio-visual
4. Print media and
5. Computer aided documentation

Assessment of ITK, setting importance and prioritization of it are essential steps for proper documentation. Besides logical assessment in the light of formal scientific theories ITK needs an informal interaction with the inhabitants of the community where the ITK is documented to elucidate relevant information pertaining to ITKs. This interaction would provide an in-depth understanding of the 'emic' perspectives of local farmers.

DOCUMENTATION OF TRADITIONAL ECOLOGICAL KNOWLEDGE FOR FISHERIES RESOURCE MANAGEMENT

Documentation with respect to gear and crafts used in the time of fishing, hydrology of the water body and small lakes, habitat preferences of fish, role of aquatic vegetation as habitat components, and impacts of habitat disturbances due to human interventions in aquatic ecosystems may be done by initiating an informal discussion with the respondents, they may be asked questions on each indigenous traditional practice of these area. Farmer participatory approach (It is a systematic, semi structured approach that uses a combination of methods to assess and understand a community's situation or a particular problem with the participation of and through the eyes of local people), focused group discussion i.e.; intensive discussion and interviewing of small groups of people on a given issue for easy documentation and arriving meaningful triangulation along with semi structured interview can be conducted for documentation of TEK. During the course of discussion, the respondents may be requested to give details of each practice and accordingly, the detailed procedure involved in following such practices need to be captured digitally and noted down in black and white in details along with the required inputs, dosage and periodicity of administration etc. The technologies will be then sorted out, based on their nature, in consultation with the experts. After completion of field survey, a comprehensive list of identified traditional technologies needs to be prepared.

Analysing the rationale of the TEK

Rationality refers to the possible scientific explanation for each of the documented indigenous technology (Nirban, 2006). Rationality refers to the degree to which TEK can be explained or supported with scientific explanations, or have been established based on long term experiences (Hussain and Sundaramari, 2011). Here a reasonable and justifiable practice or belief is considered to be rational. Rationality scale (Hiranand, 1979) may be used to judge the rationality of the indigenous practices. After collecting the traditional practices from the farmers, the practices will be listed and send to a panel of judges with around 10 years experience in the related fields for judging the scientific value of the documented TEKs and their response may be rated on 5 point score. This is a five point scale with ratings as 'Very Rational', 'Rational' ,'Undecided', 'Irrational' and 'Very Irrational' with the weightage of 5,4,3,2 and 1 respectively. The weighted mean scores of individual practices need to be then calculated. The practices and beliefs which will be assigned a weighted mean score above 3.5 may be considered to be rational. However, since many of the technologies practiced by the farmers are yet to be put to scientific test, only the views and opinions of the scientists could be collected and not the actual results of experiments. For establishing the scientific reasoning of the technologies identified through field survey, scientists' possible perception and relevant literature may be referred to find out the appropriate explanations of the technologies, wherever possible.

Phase wise study to Document the TEK for Fisheries Resource Management

- **Phase 1**

  a. Reviving reports/literature related to wetland, river environmental issues, biodiversity and habitat management of fishes

- **Phase 2**

  a. Finalize site selection through matrix ranking with the attributes such as suitable for TEK Research, extent of bio diversity reported, local level governance, number of indigenous fishermen depended on the wetland, interesting "story" for study, transportation etc.

  b. Orientation: Introductory meetings with stakeholders.

  c. Situation analysis in respect to major issues and status of selected wetland/rivers/seas with special reference to conservation

Stakeholder analysis: Stakeholder analysis is analysis of problems, expectations, interests, motivation, attitudes and potentials of all the agencies, organisations, groups and people will influence/be influenced by the project.

Tool may be used- Stakeholder analysis matrix, SWOT analysis, Venn diagram, Web diagram etc.

Problem analysis: One main tool uses in problem analysis is the 'problem tree',. Problem tree is constructed by selecting a problem from the list of problems identified and relating this problem to starter problem using the cause-effect relationship. If the problem is a cause of a starter problem it is placed below the starter problem. If the problem is an effect it must be placed above of the starter problem. If it is neither a cause nor an effect it should goes at the same level. The tree should be end up with one main problem and a series of lower order problems that branch out below the main problem.

d) Main fieldwork:

a. Transect walks and key informants/focus group discussion

b. Participatory mapping/seasonal calendars/time lines

c. Community meeting - 1 and 2 (village level)

d. Semi-structured interviews with fishers/local governance/NGO and DoF personnel

e. Detailed Investigation on selected wetlands for documentation of TEK

f. Validation of the research findings through triangulations

g. Rationality analysis of identified TEKs

- **Phase 3**

a. Organizing data, pursuing analysis, verification of the data under analysis, and technical report writing.

CONCLUSION

Globally, various efforts are underway to protect aquatic environment, aquatic bio diversity and management of natural resource on a sustainable basis. Fisher's participation, local stewardship, and shared decision-making is utmost important in the management of fisheries resources. The partnership between resource users and government has given the opportunities for using fisher ecological knowledge into modern management systems. Today, many traditional ecological knowledge systems are at risk of becoming extinct because of rapidly changing natural environments and fast pacing economic, political, and cultural changes on a global scale. The development partners need to recognize the role of TEK, understand its workings in the context of the local communities, and integrate systematically the most effective and promising of such practices into the development programs they support. The impact and sustainability of modern practices could be enhanced if they are adapted to the local conditions and the indigenous practices. Special efforts are, therefore, needed to understand, document and disseminate TEK for preservation, transfer or adoption and adaptation elsewhere. By helping to share fishers' traditional ecological knowledge within and across communities the development community can learn a lot about the aquatic conditions that affect those communities livelihood. TEK should complement, rather than compete with global knowledge systems in the implementation of projects. By investigating first what local communities know and have in terms of Traditional ecological knowledge, development partners could better help improve upon those practices by bringing to the dialogue international practices from development experiences in other parts of the world. Moreover, this process can contribute to better cross-cultural understanding and to the promotion of culture in development. But, above all, investing in the exchange of traditional ecological knowledge and its integration into the development process can help to reduce poverty by efficient handling of resources and conserving the aquatic biodiversity.

References and Recommended Readings

Bergmann, M. H., Hinz, R. R., Blyth, M. J., Kaiser, S., Roger, I. and Armstrong, M., 2004. Using knowledge from fishers and fisheries scientists to identify possible ground fish "Essential Fish Habitats". *Fisheries Research* **66**: 373-379.

Berkes, F., 1999. Sacred Ecology: Traditional Ecological Knowledge and Resource Management. Taylor and Francis, Philadelphia and London.

Berkes, F. and Folke, C., 2002. Back to the Future: Ecosystem Dynamics and Local Knowledge. In: Panarchy: Understanding Transformations in Human and Natural Systems (Gunderson, L. and C. S. Holling, eds.). Island Press, Washington, D. C., 121-146.

Colfer, C. J. P., 1990. Rice Production and Subtleties of Cultural Change: An Example from Borneo. In proceedings of International workshop on Sustainability Through Farmers Involvement in Technology Generation and Diffusion. Feb. 9, 1990. Indian Society of Agronomy. New Delhi.

Dean, D. L., 1999. Indigenous Knowledge. Technology and Blending and Gender Implications. Gender Technology and Development, New Delhi. Safe Publications. **3**(2).

Ellen, R. and Harris, H., 1996. Concepts of indigenous environmental knowledge in scientific and development studies literature - A critical assessment; draft paper East-West Environmental Linkages Network Workshop 3, Canterbury.

Eythorsson, E., 1993. Sami Fjord fishermen and the state: traditional knowledge and resource management in northern Norway. In: J. Inglis, ed. Traditional ecological knowledge: concepts and cases. Ottawa: International Programme on Traditional Ecological Knowledge/ International Development Research Centre, Canadian Museum of Nature, 133–142.

Eythorsson, E., 1998. Voices of the weak – relational aspects of local ecological knowledge in the fisheries. In: S. Jentoft, ed. Commons in a cold climate: coastal fisheries and reindeer pastoralism in north Norway: the co-management approach. Paris: UNESCO/Parthenon, 185–204.Deka, S. (n.d.). Conserving Lake Ecosystem - A case study of Rudrasagar Lake (Ramsar Site), Tripura. In Aquafind. Retrieved June 20, 2011, from http://aquafind.com/ articles/Rudrasagar_lake.php#sdfootnote1sym.

FAO (Food and Agricultural Organization), 1999. Information on fisheries management in the people's republic of Bangladesh. Available at: http://www.fao.org/fi/fcp/en/BGD/body.htm. [May 2005].

FAO (Food and Agricultural Organization), 2002. The State of World Fisheries and Aquaculture.URL:http://www.fao.org/documents/show_cdr.asp?url_file=/docrep/007/ y5600e/y5600e00.htm [March 2005].

Faulkner, A., Renato, A. and Silvano, M., 2001. Status of Research on indigenous Fishers' Fisheries. Research document 96/45, 28 p.

Gadgil, M., Olsson, P., Berkes, F. and Folke, C., 2003. Exploring the role of local ecological knowledge in ecosystem management: three case studies. In Navigating Social- Ecological System: Building Resilience for Complexity and Change (Berkes, F., J. Colding, and C. Folke, eds.), pp.189-209.

Grant, S. and Berkes, F., 2004. "One Hand Cant Clap': Combining Scientific and Local Knowledge for Improved Caribbean Fisheries Management." Presented at "The Commons in an Age of Global Transition: Challenges, Risks and Opportunities," the Tenth Conference of the International Association for the Study of Common Property, Oaxaca, Mexico, August 9-13.

Grenier, L., 1998. Working with indigenous knowledge. A guide to researchers. International

Hiranand, K. K., 1979. Technocultural profile of dryland village and dry farming technology – an international study. Ph.D. Thesis, Haryana Agricultural University, Hissar.

Husain, S. and Sundaramari, A. M., 2011. Scientific Rationality and Perceived Effectiveness of Indigenous Technical Knowledge on Coconut (*Cocos nucifera* L.) Cultivation in Kerala, *Journal of Tropical Agriculture* **49**(1-2): 78-87.

Johnson, M., 1992. Dene traditional knowledge. *Northern Perspectives* **20**(1): 3-5.

Knowledge in Australia and Brazil. In Proceedings of the conference on "Putting fishermen's knowledge to work" held at the Fisheries Centre, University of British Columbia, August 27-30, 2001, 110-116.

Mane, P. M. and Sutaria, J. N., 1993. Study and Documentation of Indigenous Knowledge/ Traditional Agricultural Practices of Tribal Farmers. Paper presented in National Seminar on Indigenous Technology for Sustainable Agriculture. 23-25 March, 1993. IARI, New Delhi, India.

McCay, B. J. and Finlayson, A. C., 1995. The political ecology of crisis and institutional change: the case of the northern cod. Paper presented at the Annual Meeting of the American Anthropological Association, Washington, 15–19 November.

McGoodwin, J. R., Neis, B. and Felt, L., 2000. Integrating fishery people and their knowledge into fisheries science and resource management: issues, prospects and problems. In: B. Neis and L. Felt, eds. Finding our sea legs: linking fishery people and their knowledge with science and management. St. John's, Newfoundland, Institute for Social and Economic Research, 249–264.

Neis, B., 1992. Fishers' ecological knowledge and stock assessment in Newfoundland. *Newfoundland Studies* **8**(2), 155–178.

Neis, B., Felt, L. Schneider, D. C., Haedrich, R. Hutchings, J. and Fischer, J., 1996. Northern cod stock assessment: what can be learned from interviewing resource users? DFO Atlantic Fisheries. Research document 96/45, 28 p.

Nirban, A. A., 2006. A Study on Indigenous Technical Knowledge about Rice Cultivation and Bovine Health Management Practices in Konkan Region of Maharashtra, Unpublished Ph.D thesis, Department of Agricultural Extensuion Education, College of Agriculture, Dharwad, University of Agricultural Sciences.

Pal, P. K., 1998. Indegenous Knowledge Base of Agro-Technology - A Study in a Lateritic Area of West Bengal. Ph. D. Thesis. Visva-Bharati, Sriniketan, West Bengal, India.

Pal, P. K. and Das Gupta, D., 1997. Participatory Rural Appraisal in identifying Intellects of Local People. (In): D. Das Gupta and S.D Mukhopadhayay (ed). Social Science on the Farm Front. Naya Prakash.

Parasar, B., 1994. Indigenous agricultural knowl¬edge of tribals in eastern ghat highland zone of Orissa. Ph.D. Thesis. BHU, Varanasi, U.P. India.

Poulsen, A. F., 2001. Integration of fishers' ecological knowledge into research on a large tropical river basin, the Mekong River in Southeast Asia. Paper presented at the conference Putting Fishers Knowledge to Work', UBC Fisheries Centre, Vancouver, BC, pp.198-207.

Quiroz, C., 1993. Methodology for the study of Farmers Agricultural Local Indigenous Knowledge System in Rural Development Programme. Inservice training programme for research/ extension practitioner—an experience from Venezuela. *Indigenous Knowledge and Development Monitor* **1**(2).

Rajshekaran, B., 1993. A framework for incorporating indigenous knowledge system into agricultural research, extension and NGOs for sustainable agricultural development. Studies on technologies and social change, No. 21. Technologies and social change programme, IOWA, State University, Ames, IOWA, 5001, USA.

Sillitoe, P., 2000. State of indigenous knowledge in Bangladesh. In: Indigenous Knowledge Development in Bangladesh: Present and Future, ed. The University Press Limited, Dhaka.

Thomson, P. M., 2003. Understanding livelihoods dependent on inland fisheries in Bangladesh and Southeast Asia. DFID/FMSP Project R8118. Bangladesh Summary report. Retrieved from: http://www.ahj00.dial.pipex.com/r8118/r8118_13. pdf [Accessed May 2005].

UNEP/CBD/TKBD, 1997. Final Document of the Second International Indigenous Forum on Biodiversity, UNEP/CBD/TKBD/1/3 Annex 1.

World Bank, 1998. Indigenous Knowledge for Development-A Framework for Action, Knowledge and Learning Center Africa Region.

Chapter – 4

Indigenous Technical Knowledge for the Management of Major Pests of Field Crops in North East India

D. M. Firake, P. D. Firake , G. T. Behere, N.S. Azad Thakur and Dilip Nath

INTRODUCTION

North East India, a mega-biodiversity centre and a hotspot (Myer et al., 2000) occupy 7.7% of India's total geographical area supporting 50% of the flora (ca. 8000 species) (Rao, 1994), of which 31.58% (ca. 2526 species) are endemic (Nayar, 1996). It is a transitional zone between the Indo-China and Indo-Burma–Malaysian regions. This region is also a part of the vavilovian centre of biodiversity and origin of many important cultivated plant species and some domesticated animals (Agarwal, 1996). North Eastern Hilly (NEH) region of India supports a rich biodiversity spanning from tropical rainforests to alpine scrubs. Takhtajan (1969) described the region as the "Cradle of Flowering Plants" because of its diversified angiosperms. This rich biodiversity has a significant role to balance the ecosystem and has been used ethnologically by locals for various socio-economic and developmental purposes (Yumnam, 2008).

According to Biodiversity Strategy and Action Plan for North East Eco-region, 3,624 species of insects and 50 molluscs are found in the region (http://himalayanconnections. org/wp-content/uploads/pdf/Chatterjee_2006_ BiodiversitySignificanceNEIndia_WWF.).

Greater part of the agriculture of this region is comes under hill farming and is comparatively more vulnerable to insect pest infestation due to the favourable climatic conditions. Specificity of locations in hills is determined by factors like altitudinal differences, steepness of slopes, directions of mountain and valley etc. Microclimatic specificity of some hilly pockets is becoming hot spot for certain pests. The agro climatic conditions of the region are very conducive for the growth and multiplication of the insect fauna and their natural enemies and they pose serious problems with resultant low productivity in almost all the crops. Insect pests inflict enormous losses to the agricultural production. Common insect pest species found in important field crops are presented in Table 1 to 4 (rice, maize, pulses and oilseeds, respectively).

Table 1: Major insect pests of rice in Meghalaya

Insect pests	Scientific name	Family	Order
Yellow stem borer	*Scirpophaga incertulus*	Crambidae	Lepidoptera
Leaf folder	*Cnaphalocrosis medinalis*	Pyrostidae	Lepidoptera
Case worm	*Nimphula depunctalis*	Pyrostidae	Lepidoptera
Hispa	*Dicladispa armigera*	Hispidae	Coleoptera
Gundhi bug	*Leptocorisa oratorius*	Alydidae	Hemiptera
Swarming caterpillar	*Spodoptera mauritia*	Noctuidae	Lepidoptera
Thrips	*Thrips oryzae*	Thripidae	Thysanoptera
Gall midge	*Orseolia oryzae*	Cecidomyidae	Diptera
Army worm	*Mythiman separata*	Noctuidae	Lepidoptera
Root aphids	*Rhaphalosiphum rufiabdomanalis* *Tetraneura nigriabdomanalis*	Aphididae	Hemiptera

(*Source*: Anonymous, 2006 and Thakur et al., 2009)

Table 2: Major insect pests of maize

Pest	Scientific name	Family	Order
Cob borer	*Stenachroia elongella*	Cosmopterigidae	Lepidoptera
Stem borer	*Chilo partellus*	Crambidae	Lepidoptera
Pink borer	*Sesamia inferens*	Noctuidae	Lepidoptera
Shoot fly	*Atherigona soccataA. falcata*	Muscidae	Diptera
Aphids	*Hysteroneura setariae* *Rhopalosiphum maidis* *Popillia cupricollis*	Aphididae	Hemiptera
Green bugs	*Dolycoris indicusNezara viridula*	Pentatomidae	Hemiptera
Cutworm	*Agrotis flammatra A. segetum*	Noctuidae	Lepidoptera
Grasshopper	*Hieroglyphus banian* *Atractomorpha crenulata* *Chrotogonus* sp. *Acrida exaltata*	Acrididae	Orthoptera
Termites	*Odentotermes* spp. *Microtermes* spp.	Termitidae	Isoptera

(*Source*: Anonymous, 2006 and Thakur et al., 2009)

Table 3: Major insect pest of pulses

Pest	Scientific name	Family	Order
Pod boring weevil	*Apion clavipes*	Curculionidae	Coleoptera
Blister beetles	*Mylabris pustulataMylabris phalerata*	Meloidae	Coleoptera
Pod borer	*Catechrysops cnejus*	Pyrostidae	Lepidoptera
	Helicoverpa armigera	Noctuidae	
	Maruca testulalis	Noctuidae	
Tur Pod fly	*Melanoagromyza obtusa*	Agromyzidae	Diptera
Tur plume moth	*Exelastis atomosa*	Pterophoridae	Lepidoptera
Pea aphid	*Aphis craccivora*	Aphididae	Hemiptera
Mites	*Aceria cajani*	riophyidae	Acarina

(*Source*: Anonymous, 2006 and Thakur et al., 2009)

Table 4: Important insect pests of oilseeds

Pest	Scientific name	Family	Order
Leaf folder	*Nacoleia vulgalis* and *N. diemenalis*	Pyrostidae	Lepidoptera
Stem fly	*Ophiomyia phaseoli*	Agromyzidae	Diptera
Hadda beetle	*Epilachna vigintioctopunctata* *E. varivestisCerotoma* spp.	Coccinellidae	Coleoptera
Leaf miner	*Aproaerema modicella*	Agromyzidae	Diptera
White grub	*Holotrichia serrata,* *H. consanguinea* *Leucopholis lepidophora*	Melolonthidae	Coleoptera
Aphids	*Lipaphis erysimi* *Brevicorne brassicae* *Myzus persicae*	Aphididae	Hemiptera
Saw fly	*Athalia lugens proxima*	Tenthridinidae	
Leaf webber	*Antigastra catalaunalis*	Pyrostidae	Lepidoptera

(*Source*: Anonymous, 2006 and Thakur et al., 2009)

INDIGENOUS TECHNICAL KNOWLEDGE (ITK) AND PEST MANAGEMENT

ITK is the knowledge of the indigenous people spread over different geographical regions of the world with their own language based on their culture, tradition, belief, folklore, rites and rituals. During the course of their close association with nature and natural resources, they have invented various solutions or remedies to their problems which they have encountered in their day to day life while managing the land and environmental resources for their survival (Chhetri and Belbahri, 2009). Recently, the organic agriculture is emphasized, particularly in rich biodiversity areas like North East India (Deka et al., 2006). ITK is distinct from the international knowledge, which is generally derived through hypotheticoinductive process such as knowledge generated in universities, research organizations, private, and public research institutions through research activities (Chhetri and Belbahri, 2009).

The farmers from different ethnic diversity of the region have developed their own systems of cultivation of various crops (Sharma and Gogoi, 1999). They have established distinct system of knowledge, innovation and practices relating to the uses and management of biodiversity on the land and environment. Land use systems of North East India are dominated by jhum farming where management of crops and pests are carried out by integrating indigenous knowledge and traditional ecological knowledge of the communities. This fusion of knowledge has potential and could serve as remedies for various deficiencies of pest management in modern agriculture (Chhetri and Belbahri, 2009). Uses of non-chemical methods of pest control are already gaining importance and are becoming popular in several countries including India.

The IPM (Integrated Pest Management) strategies developed and promoted by the Governments are now based on these traditional practices for better management. If more attention is given towards the production or promotion of ITK based products on cottage scale, it could be an economically viable option for sustainable development of eco-friendly pesticides/insecticides. Moreover, due to rapid urbanization, the ethnic groups are promptly reducing their conservative attitude for which the long earned ITK that they are carrying from generation to generation are undocumented and quickly disappearing (Deka et al., 2006). In this context, collection, compilation and scientific evaluation of the ITKs are very important in terms to strengthen modern agricultural research and biodiversity conservation.

Some traditional pest management practices have been documented from the region in various forms but still there are several practices, which are yet to be documented. In many cases, these practices are found to be very efficient for management of most notorious and uncontrollable pests. Furthermore, all the documented practices are largely scattered and not easily accessible. Therefore, there is an urgent need of compilation of all these practices together as one document; so as to get access and necessary help to the farmers of the region, extension workers and scientific community all over the world.

Anecdotal evidences also indicates rise in losses, despite increasing use of chemical pesticides. Furthermore, comparatively sustainable and safer chemical as well as biological pesticides are not available to the farmers residing in remote places. However, under such circumstances, only traditional practices are playing

Table 5. Traditional practices for the management of crop pests in north east India

Crop	Insect pests	ITK practices	Belief/Remark	References
Rice	Stem borer	Erection of bamboo (*Bambusa indica*) branches or other stick in the rice field.	Attract birds to take rest on the bamboo stick. The birds feed the adult moths of stem borer. Additionally sap of raw bamboo shoot contain hydrocyanic acid and possesses antiseptic and larvicidal properties (Anonymous, 1948).	Deka et al., 2006
Rice	Stem borer	Steam decoction of neem (*Azadirachta indica* A. Juss.) seeds and leaves, and spraying of extract onto to the rice crop. The extract is prepared by mixing 1 to 3 gm of ground neem seed or leaf in 1L of water (0.1-0.3% concentration) for 12 hrs (Bhogayati, 2001).	The azadirachtin present in the neem seed and leaf acts as a antifeedant and growth retardant to insect (Ley Steven, 1994)	Deka et al., 2006
Rice	Stem borer	Keeping slices of pumalo (*Citrus grandis* Osbeck) in the paddy field @1 trap/6 m^2.	The essential oils of pumalo repel stem borer (Pathak et al., 2001).	Deka et al., 2006
Rice	Stem borer	The grounded pulp of the Khira leaf introduction in irrigation channel	Repellent	Sharma et al., 2009
Rice	Stem borer	Placing chopped leaves of Indian Rhododendron or Phutuka (*Melastoma malabathricum* Linn.) in paddy field	The chopped leaves emit foul smell, which drive away the insect from field	Deka et al., 2006
Rice	Stem borer	Placing of grounded bark of drumstick (*Moringa oleifera* Lam.) in to the rice field.	Bark may contain insecticidal principles. Drumstick has been reported to have medicinal properties (Vaidyaratnam, 1995).	Deka et al., 2006

(*Contd.*)

Crop	Insect pests	ITK practices	Belief/Remark	References
Rice	Stem borer	Placing few branches of fern (*Cybotium* spp.).	Fern has insecticidal activity (Bhuyan, 2003).	Deka et al., 2006
Rice	Stem borer	Burying the puthi or barb fish (*Puntius* spp.) in to the soil for 15-20 days and spraying water extract of them in to the rice field	The fermented fish extract repels stem borer.	Deka et al., 2006
Rice	Stem borer	Placing citrus or Sakala tenga (*Citrus sinensis* Osbeck) peels in the rice field	Repellent	Deka et al., 2006
Rice	Stem borer	Erecting or pegging branches of*Cymbopogon khasianum or Saccharum spantaneum*	Growth inhibition/killing action	Chhetri and Belbahri, 2009
Rice	Stem borer	Spreading of grounded pulp of *Khira* leaves in water	Repellent	Sharma et al., 2009
Rice	Stem and root borers	Placing of branches of *Calotropis procera* in paddy fields.	Repellent	Shrivastava et al., 2009
Rice	Leaf folder	Removal of grasses around the bund of paddy field	Alternate hosts destruction	Sharma et al., 2009
Rice	Leaf folder	Placing of well fermented wine pomace (wine residue) usually made up of millets in the source of irrigation canal of rice fields	Repellent against leaf folder and also growth inhibition of rice blast fungus	Chhetri and Belbahri, 2009
Rice	Rice thrips	Spraying of aqueous suspension of cow dung in the nursery	Repellent	Chhetri and Belbahri, 2009
Rice	Case worm	Dragging of rope impregnated with kerosene in standing water	Toxicity of kerosene	Chhetri and Belbahri, 2009
Rice	Rice hispa	Rearing duck near the paddy field	Predation	Chhetri and Belbahri, 2009
Rice	Rice hispa	Mixing cow dung with water in paddy field.	Repellent	Chhetri and Belbahri, 2009

(*Contd.*)

Crop	Insect pests	ITK practices	Belief/Remark	References
Rice	Rice hispa	Placing of Nishinda plant twigs (*Vitex negundo* L.) in the infected rice fields pest	The foul smell of the plant repels the	Bhattacharjee and Ray, 2010
Rice	Gundhi bug	Burning of bicycle tyres near the rice field before panicle initiation.	Repellent	Chhetri and Belbahri, 2009
Rice	Gundhi bug	Placing rotten crab or frog in plastic traps @ 1 trap/10m^2 area	Attractant	Chhetri and Belbahri, 2009
Rice	Gundhi bug	Utongthangmei (burning of clothes on the bamboo sticks in the four corners).	Insects get attracted towards fire and subsequently die.	Bhattacharjee and Ray, 2010
Rice	Gundhi bug	A rotten crab is placed on a stick in a infected field	Attractant	Bhattacharjee and Ray, 2010
Rice	Gundhi bug	Placing Kuthap (*Clerodendrum viscosum* Vent.) twigs randomly in the infected field	Repellent	Bhattacharjee and Ray, 2010
Rice	Aphids	Placing the leaves of *Calotropis procera* in paddy fields. Placing of long branches along with leaves at a distance of 14.6 m in paddy fields; around 25-30 branches are required for 0.24 ha.	Attractant	Shrivastava et al., 2009
Rice	Brown plant hopper	Placing grounded oak tree bark in the source of irrigation canal	Repellent as well as killing action	Chhetri and Belbahri, 2009
Rice	Brown plant hopper	Burying leaves of *Calotropis procera* in the soil deep in nursery as well as in rice field.	Repellent	Shrivastava et al., 2009
Rice	Multiple pests	Spreading leaves of *Artemisiavulgaris, Croton caudatus, Munromia wallichi, Adhatoda vessica etc.*	Decomposed plant products have repellent and killing action	Chhetri and Belbahri, 2009

(*Contd.*)

Crop	Insect pests	ITK practices	Belief/Remark	References
Rice	Multiple pests	Intercropping of maize in jhum rice	Intercrops provide physical barriers Acting as a bird pearchesMaize helps in locating the burrows of rodents in paddy field	Chhetri and Belbahri, 2009
Rice	Grasshopper	Keeping tree boles/trunk and partially cut alder trees amidst jhum field	Predation: it facilitates perching of diverse birds for selective prey of pest such as grasshoppers of paddy,	Chhetri and Belbahri, 2009
Rice	Multiple pests	Growing of intercrops such as maize and sorghum long duration pigeon pea etc. with jhum rice.	Act as a birds perches	Chhetri and Belbahri, 2009
Rice	Multiple pests	Conservation of sacred grooves	Carnivorous birds prey on paddy pestsFacilitate perching of birds Excreta of these birds acts as fumigant to the growth of pathogens	Chhetri and Belbahri, 2009
Rice	Multiple pests	Indigenous repeated ploughing technique for getting rid of soil borne insects and their diapausing stages which damage paddy crops is very effective in wet land system and dry sedentary farming system (Manipur)	Predation of different life stages of insects in the soil	Chhetri and Belbahri, 2009
Rice	Multiple pests	Keeping nitrogen fixing legume viz., nepalese alder, *Alnus nepales* in jhum rice field	Enrichment of soil as well as act as a bird perches.Excreta of these wild birds act as pesticides and fungicides for crops	Chhetri and Belbahri, 2009
Rice	Multiple pests	Lightning earthen ware lamps	Repellent	Barooah and Pathak, 2009
Maize	Pest and fungal attack in hanging maize cobs	*Chang-Ghar'* of *Lisus* Tribe of Arunachal Pradesh:House built on platform of bamboo or wooden planks with a kitchen in a centre	This structure of house promote equal distribution of smoke	Sarma et al., 2006

(*Contd.*)

Crop	Insect pests	ITK practices	Belief/Remark	References
Wheat and Maize	Stem borer, *Chillo zonellis*	Spreading of grounded pulp of the Khira leaf in the wheat field	Repellent and the scent of the pulp is sufficient to kill the pests	Sharma et al., 2009
Wheat and Maize	Termite	Dipping of sugarcane setts in *Calotropis procera* leaf extract @ 15% and 20%	Antifeedant	Shrivastava et al., 2009
Wheat and Maize	Termite	Pouring *Calotropis procera* leaf extract solution in termite-infested soil	Antifeedent	Shrivastava et al., 2009
Wheat and Maize	Army worm	Spreading *Calotropis procera* leaves on the standing crops. Collect and destroy the insects gathered on the broadcasted leaves and simult aneously, replace with the fresh leaves	Attractant	Shrivastava et al., 2009
Cotton	Bollworm and other defoliators	Spraying of the latex of *Calotropis gigantea* roots, branches and leaves, diluted in 15 parts of water	Antifeedent	Shrivastava et al., 2009
Groundnut	Red hairy caterpillar	Spreading of *Calotropis gigantea* twigs on field boundaries	Repellent	Shrivastava et al., 2009
Castor and Sesamum	Red hairy caterpillar	Spreading of *Calotropis* twigs in the field	Attractant: Plant twigs attract red hairy caterpillars. Smell of leaf latex attracts caterpillars	Shrivastava et al., 2009
Mustard	Saw fly	Application of *Calotropis* leaf powder @ 50 kg/ha	Repellent	Shrivastava et al., 2009
Pigeon pea	Root knot nematode	Soil application of *Calotropis* leaf powder @ 100 kg/ha	Repellent	Shrivastava et al., 2009

considerable role in pest management programs. Therefore, these practices are found to be blessing given by their ancestors. By considering the importance these practices, we tried to compile the available research work in the form of this chapter. The important traditional pest management practices are presented in Table 5.

FUTURE PROSPECTS

Many ITKs in the region are found to be really efficient against the management of crop pests due to their higher cost, less effective than traditional practices and even some times found harmful for the environment. These practices are eco-friendly and easy to be followed by the farmers. However, some ITK practices are comparatively less effective, harmful for the environment and relatively expensive as well. Therefore, efficient traditional practices are needs to be proved scientifically, so that many products would be developed commercially and could be use as one of the component for effective and sustainable pest management.

SUMMARY

- Farmers belongs to different ethnic diversity of the North Eastern Hill region of India have developed their own systems of cultivation of various crops.
- They have established distinct system of knowledge, innovation and practices relating to the uses and management of biodiversity on the land and environment.
- Land use systems of North East India are dominated by jhum farming where management of crops and pests are carried out by integrating indigenous and traditional ecological knowledge of the communities.
- This fusion of knowledge has potential and could serve as remedies for various deficiencies of pest management in modern agriculture.
- These traditional non-chemical methods of pest control are gaining importance and are becoming popular in several countries including India.
- Many traditional practices from this region are documented, but these are largely scattered and not easily accessible.
- Therefore, attempts are made in this book chapter to compile these practices together as one document; so as to get access and necessary help to the farmers of the region, extension workers and scientific community all over the world.

References and Recommended Readings

Agarwal, K. C., 1996. *Biodiversity*. Agra Botanical Publishers, India. 118 p.

Anonymous, 1948. The Wealth of India, Raw Materials, Vol 1, (Council of Scientific & Industrial Research, New Delhi). 254.

Anonymous, 2006. Integrated management of insect pest of crops in north eastern hill region. Technical Bulletin No. 19. ICAR RC for NEH Region, Umiam, 50p.

Barooah, M. amd Pathak, A., 2009. Indigenous knowledge and practices of *Thengal Kachari* women in sustainable management of *bari* system of farming. *Indian Journal of Technical Knowledge* **8**(1): 25-40

Bhagavati, R. R., 2001. Home made agricultural medicine (Assamese). *Mrittika*. **3**(2): 43-45.

Bhattacharjee, P. P. and Ray, D. C., 2010. Pest management beliefs and practices of Manipuri rice farmers in Barak valley of Assam. *Indian Journal of Traditional Knowledge* **9**(4): 673-676.

Bhuyan, M., 2003. Studies on some potential insect control agents from plants of North East India, PhD Thesis, Dibrugarh University, Dibrugarh, Assam.

Chhetry, G. K. N., Belbahri, L., 2009. Indigenous pest and disease management practices in traditional farming systems in north east India: A review. *Journal of Plant Breeding Crop Sciences* **1**: 028-038.

Deka, M. K., Bhuyan, M. and Hazarika, L. K., 2006. Traditional pest management practices of Assam. *Indian Journal of Traditional Knowledge* **5**: 75-78.

Ley Steven, V., 1994. Synthesis and chemistry of the insect antifeedant azadirachtin. *Pure Applied Chemistry* **66**(10/11): 2099-2102.

Myer, N., Muttermeier, R. A., Muttermeier, C. A., da Fornseca, G. A. B. and Kent, J., 2000. *Nature* **403**: 853–858.

Nayar, M. P., 1996. Hotspots of Endemic Plants of India, Nepal and Bhutan. SB Press, Trivandrum.

Pathak, K. A., Thakur, N. S. A., Rao, K. R. and Shylesha, A. N., 2001. Insect pests of crops and their management. In Steps towards modernization of agriculture in NEH region, *ed* Verma, N. D. and Bhatt, B. P. (Indian Council of Agriculture Research, New Delhi), 121-159.

Rao, R. R., 1994. Biodiversity in India: Floristic Aspects, Publisher Bishen Singh Mahendra Pal Singh, Dehra Dun.

Sarma, R., Arunachalam, A., Adhikari, D. and Majumdar, M., 2006. Indigenous technical knowledge and resourse utilization of *Lisus* in the south eastern part of Namdapa National Park, Arunachal Pradesh. *Indian Journal of Traditional Knowledge* **5**(1): 51-56.

Sharma, K. K. and Gogoi, R., 1999. Agricultural traditional wisdom in Assam. *Assam Agri-history* **3**(3): 199-206.

Sharma, S., Bajaracharya, R. and Sitaula, B., 2009. Indigeneous Technology Knowledge.

Srivastava, S. K., Babu, N. and Pandey, H., 2009. Traditional insect bio-prospecting- as human food and medicine. *Indian Journal of Traditional Knowledge* **8**: 485-494.

Takhtajan, A., 1969. Flowering Plants: Origin and Dispersal, Oliver and Boyd Ltd., Edinburgh.

Thakur, N. S. A., Kalaishekhar, A., Ngachan, S. V., Saikia, K., Rahaman, Z. and Sharma, S., 2009. Insect pest of crops in North East India. ICAR Research Complex for NEH region, Umiam.

Vaidyaratnam, P. S., 1995. Indian Medicinal Plants, 4, Varier's Arya Vaidya Sala, (Orient Longman Ltd).

Yumnam, J. Y., 2008. Rich biodiversity of North East India needs conservation. *Current Science* **95**(3): 297p.

Online sources

http://himalayanconnections.org/wp-content/uploads/pdf/Chatterjee_ 2006_ Biodiversity SignificanceNEIndia_WWF (as on 19. 02. 13).

Chapter – 5

Traditional Practices for the Management of Major Pests of Horticultural Crops and Stored Produce in North East India

D.M. Firake, P.D. Firake, G.T. Behere, N.S. Azad Thakur and D. Nath

INTRODUCTION

The North Eastern Hill (NEH) Region of India comprising the states of Arunachal Pradesh, Assam, Meghalaya, Mizoram, Manipur, Nagaland, Tripura and Sikkim lies between 20°5' to 29°30' latitude and 89°49' E to 97°3' E longitude. The region possesses a great diversity in soil and climate as influenced by varied altitudes ranging from 18m MSL to more than 3000m MSL. In the entire region due to high rainfall (1800 mm to 3000 mm/year) humidity is very high (50-100%). The average minimum and maximum temperature is also comparatively low and ranges from below 0°C to a maximum of 35°C. This region is very rich in natural resources with diverse flora and fauna.

The main horticultural crops in this region are vegetables (beans and peas, brinjal, bhendi, cole crops, cucurbits, onion and tomato), fruit crops (banana, citrus, guava, mango, apple, peach, pear and plum), tuber crops (colocasia, potato, sweet potato and yams) and spices (chilies, ginger, large cardamom, and turmeric). As the climatic conditions of the region are conducive for different types of horticultural crops but at the same time these conditions are also favorable for the occurrence and multiplication of insect pests and their natural enemies. These insect pests cause colossal damage to crops resulting into low production in the region. Moreover, stored grain pests including rodents cause huge losses in every part of the world and including NEH region where most wondering phenomenon of nature 'gregarious flowering of bamboos' is very common. Almost all the storage commodities are not escape from the damage of these notorious pests. However, some tribal practices are found to be excellent for the management of these pests.

The production of horticultural crop is being increasingly given more and more importance by the farmers due to the high yield and demand in the market. The farmers tend to practice monocropping with high yielding varieties and hybrids to increase the production. This monocropping results into high insect pest prevalence ultimately forcing the farmers to dump extremely high amount of chemical pesticides onto their crops to control these pests, which is more often unwanted. Apart from increasing the cost of production, it also poses great health hazard to the consumers. Main reason behind the over-use of pesticides is lack of proper knowledge on pest identity.

The farmers from different ethnic diversity of the region have developed their own systems of cultivation of various crops known as 'Traditional Knowledge' (TK). The use of non-chemical methods for pest control and crop protection is already gaining importance in several countries including India. Most of this traditional knowledge is published in various reports and still several things are yet to be published from many corners of the country. However, there is lack of comprehensive information at one place. Therefore, we tried to summarize all these available techniques in this chapter. The important insect pests of the horticultural crops in the region are listed in Table 1 to 7. The common traditional practices followed by tribal farmers of the region and their general beliefs are presented in Table 8.

Table1: Major insect pests of cole crops

Common name	Scientific name	Nature of damage
Cabbage butterfly	*Pieris brassicae*	Defoliation
Dimond back moth	*Plutella xylostella*	Skeletonisation and defoliation
Cabbage flea beetle	*Plyllotreta cruciferae*	Shot holes on leaves, buds and stem
Cabbage aphid	*Brevicoryne brassicae*	Yellowing, curling, drying and sooty mould formation
Painted bug	*Bagrada cruciferarum*	Yellowing and drying of leaves
Leaf caterpillar	*Spodoptera litura*	Defoliation

Table 2: Major insect pests of solanaceous crops

Common name	Scientific name	Nature of damage
Brinjal shoot and fruit borer	*Leucinodes orbanalis*	Drying and drooping of shoot and holes on fruit
Brinjal leaf roller	*Eublemma olivacea*	Leaf rolling
Hadda beetle	*H. vigintioctopunctata* *E. dodecastigma*	Leaf skeletonisation, browning and drying
Mealy bug	*Coccidohystrix insolita*	Yellowing, wilting and drying of leaves
Tomato fruit borer	*Helicoverpa armigera*	Holes on fruit
Patato aphid	*Myzus persicae*	Yellowing and drying of leaves

Table 3: Major insect pests of Bhendi

Common name	Scientific name	Nature of damage
White fly	*Bemisia tabaci*	Curling, premature falling of flowers and fruits, yellow vein mosaic
Dusky cotton bug	*Oxycarenus laetu*	Shriveled fruits
Red cotton bug	*Dysdercus cingulatus*	Shriveled fruits
Spotted fruit borer	*Earias vittella*	Fruit damage
Blister beetle	*Mylabris pustulata*	Flower and buds feeder

Table 4: Major insect pests of cucurbitaceous crops

Common name	Scientific name	Nature of damage
Hadda beetle	*H. vigintioctopunctata* *E. dodecastigma*	Leaf skeletonisation, browning and drying
Fruit fly	*Bactrocera cucurbitae B. tau*	Fruit damage

Table 5: Major insect pests of fruit crops

Common name	Scientific name	Nature of damage
Citrus Trunk borer	*Anoplophora versteegi*	Trunk tunneling, drying of branches
Citrus leaf miner	*Phyllocnistis citrella*	Leaf mining, curling and defoliation
Lemon butterfly	*Papilio demoleus*	Defoliation
Citrus psylla	*Diaphorina citri*	Wilting, drying and vitality of plant affected
Citrus black fly	*Aleurocanthus woglumi*	Curling, premature falling of buds and fruits
Guava trunk borer	*Aristobia testudo*	Withering and drying of branches
Guava fruit fly	*Bactrocera dorsalis*	Rotten fruits, unfit for consumption

Table 6: Major insect pest of stored grains

Commodity	Insect Pests
Paddy	Rice weevil *(Sitophilus oryzae)* Rice Moth *(Corcyra cephalonica)* Rust Red flour beetle *Tribolium castaneum,* Angumois Grain Moth *(Sitotroga cerealla)*
Maize	Maize weevil (*S. zeamays*) Angumois Grain Moth *(Sitotroga cerealla)* Meal Moth *(Plodia interpunctella)* *Carphophillus* spp
Pulses	Pulse beetle *(Callosobruchus* spp.*)*

Table 7: Rodent pest of crops in North East India

Common name	Scientific name
Lesser Indian mole rat	*Bandicota bengalenis* (Gray)
Common house rat	*Rattus rattus* (Linnaeus)
Norway rat	*Rattus norvegicus* (Berkenhout)
Himalayan rat	*Rattus nitidus* (Hodgson)
Sikkim rat	*Rattus sikkimensis* (Hinton)
Mackenzie's rat	*Berlylmys mackenziei* (Thomas)
Bay bamboo rat	*Cannomys badius* (Hodgson)
Himalayan chestnut rat	*Niniventer fluvescens* (Thomas)
House mouse	*Mus musculus* (Waterhouse)
Bower's rat	*Berylyms bowersii* (Anderson)
Irrawady squirrel	*Callosciurus pygerithrus* (Geoffroy)

Table 8: Traditional practices for the management of major pests of horticultural crops and stored produce in North-East India

Crop	Insect pests	Traditional practices	Belief/Remark	Reference
Vegetables	Sucking pests	Sprinkling wood ash on vegetable crops	Repellent	Barooah and Pathak, 2009
Vegetables	Many pests	Growing border mixed crops (broad leaf types) such as pumpkin, cucumber, lady's finger, banana and other horticulturalcrops like guava, pineapple, etc.	Attractant and killing	Chanu, 2010
Vegetables (Kitchen garden)	Bugs and beetles	Local crabs are smashed and put on the top of a pointed stick and is placed in crop fields	Attractant and killing	Sinha et al., 2004
Vegetables (Kitchen garden)	Defoliators	Spraying of mixture of Cowdung, Cowurine, Chili & Garlic @ 2:1:0.5:0.25 ratio	Killing	Sinha et al., 2004
Vegetables	Defoliators	Spraying of dissolved silkworm excreta	Repellent	Sinha et al., 2004
Vegetables	Lepidopteran larva	Spraying of "Re'not-bol" plant extract	Killing	Sinha et al., 2004
Fruit trees	Many pests	Fruit trees (*Bridelia retusa*) and animal bone tied with horticultural plants	Attractants for attracting pest predators	Saravanan, 2010
Kitchen garden and fruit trees	Lepidopteran pests	Use of 'Changsim' (*Sapium baccatum* Roxb.) or 'Tuthekme' [*Dendropthoe falcata* (L.) Elting.] or 'Rakseng' (*Morus macroura* Mig.) or 'Khasi-bol' (*Bridelia retusa* Spreng.) fruits to attract insect predators (birds).	Attractant for predators	Sinha et al., 2004
Ginger	Many pests	Planting of *Calotropis* in the field of ginger	Repellent	Shrivastava et al., 2009
Ginger	Many pests	Bhang leaves and soil is mixed in the ratio 1:1 and is kept for 12-24 hours for proper intermingling.	Repellent	Sinha et al., 2004
Pumpkin	Fruit fly	Smoking in the pumpkin field	Repellent	Deka et al., 2006
Potato	Tuber moth	Sand cover (2-5 cm depth) over the stored potato.	Female moths can	Deka et al., 2006

(Contd.)

Crop	Insect pests	Traditional practices	Belief/Remark	Reference
			not lay eggs on the surface	
Citrus	Defoliators	Placing of red tree ant (*Oecophylla smaragdina*) nest on the citrus plant.	Predation	Deka et al., 2006
Citrus	Citrus trunk borer	Application of fish cleaned water at the base of plant	Repellent	Barooah and Pathak, 2009
Citrus	Citrus trunk borer	Application of fish cleaned water at the base of plant	Fishy water attracts red ants, *Oecophylla smaragdina*	Deka et al., 2006
Citrus	Citrus trunk borer	Introducing predacious red tree ants nest into the fruit orchard	Repellent	Barooah and Pathak, 2009
Citrus	Citrus trunk borer	Application of kerosene oil to the fruit tree trunks	Repellent	Barooah and Pathak, 2009
Banana	Snails and slugs	Common salt is applied to the base of plants	Repellent	Barooah and Pathak, 2009
Jack fruit, mango	Fruit moth	Smoke is generated at the base of fruit trees	Prevent pest infestation of fruits	Barooah and Pathak, 2009
Fruit orchard	Birds and monkeys	Catapults and drum beating	Wards off birds and monkeys, that swarm the homestead during period of fruiting and crop maturity	Barooah and Pathak, 2009
Arecanut	Red palm weevil	*'Peit Ksain Kwai'*(Checking the grubs in nuts) Killing the larva by inserting wire hook in the bored hole and also form soil below the tree	Mechanical control	Umdor, 2004
Coconut	Rhinoceros beetle	Placing of long hair of women in the crown portion of coconut tree	The beetle engulfed with long hair due to which they cannot move	Deka et al., 2006
Coconut	Rhinoceros beetle	Placing of a dead frog at the base of the coconut	Repellent	Deka et al., 2006

(*Contd.*)

Crop	Insect pests	Traditional practices	Belief/Remark	Reference
		plant @ 1 trap/m2.		
Coconut	Squirrel	The fragmented human hair and dry fish mixture are kept in the crown	Along with the fish the fragmented hairs are also consumed, which produces adverse effects to the squirrel.	Deka et al., 2006
Stored grains	Rice weevil	Keeping Heigri (*Dillenia indica* L.) leaves over the stored rice	Repellent	Bhattacharjee and Ray, 2010
Stored grains	Rice weevil	Keeping neem leaves and branches over the stored rice	Repellent	Bhattacharjee and Ray, 2010
Stored grains	Rice weevil	Placing of Tezpata, (*Cinnamomum tamala* Nees & Eber m.) leaves on storage grains	Repellent	Bhattacharjee and Ray, 2010
Stored grains	Rice moth	Keeping of Chawai sabi (*Polygonum hydropiper* Linn.) leaves over the stored rice	Repellent	Bhattacharjee and Ray, 2010
Stored grains	Many pests	*Artemisia vulgaris* L. (Maharna). Put fresh or dried branches/leaves in and around granaries	Repellent Insecticidal (Duke et al., 2002)	Sinha, 2010
Stored grains	Many pests	*Azadirachta indica* (Neem), *Cympopogan khasianus* (Shipuh harvashe) and *Cannabis sativa*: Crush and put fresh or dried branches/ leaves in and around granaries	Repellent *A. indica* (Chopra et al., 1941), *C. khasianus* contain Geraniol (Choudhury and Leclercq, 1996), *C. sativa* (Blackwell, 1990)	Sinha, 2010
Stored grains	Many pests	*Elshotzia blanda* (kholo) or *Melia composita* (Thoso): crush and put fresh and dried leaves in and around granaries	Repellent (Rao, 1981) *E. blanda* Active principle: Leteolin and Friedelin (Chen et al., 2005) Insecticidal: *M. composite* (Anonymous, 1997)	Sinha, 2010

(*Contd.*)

Crop	Insect pests	Traditional practices	Belief/Remark	Reference
Stored grains	Many pests	Rub pulses with pork oil before storage	Repellent	Sinha, 2010
Stored grains	Many pests	Kitchen storage: Pulse, maize, onion etc. are kept in the kitchen around 2-3 m above the ground at an angle of 450 from the chullah (cooking place)	Kitchen smoke drives away insects and reduces fungal attack	Sinha, 2010
Stored grains	Many pests	Indigenous granary : Harvested grains are stored in special granaries, constructed outside the house for proper aeration and tin sheets or holed wooden plates are provided as rat barriers in the pillars	It prevents the entry of insect pests and rats into the granary	Sinha, 2010
Stored grains	Rodents	Gilmat (*Dendrocnide sinnuata*): Place fresh leaves in between two layers of boiled rice. After 24 hrs, put treated rice in possible rat runways	Rodenticide : Leaves contain a toxic substance (Formic acid)	Sinha et al., 2004 and 2010
Stored grains	Rodents	*Entada pursaetha* (kah and sui): mix equal proportion with groundnut seed kernel and ground rice and put in possible rat runways	Repellent (Lalramnghinglava, 1996), Rodenticide, Active principle: Triterpenoid and Saponin (Tapondjou et al., 2005)	Sinha et al., 2004 and 2010
Stored grains	Rodents	Plugging of burrows by paddy husk and local chilly land races	The suffocating pungent smokes promptly affect the respiration systems of rats and killed.	Chhetry, 2009
Stored grains	Rodents	'Rat Proof Granary' : '*Nahu*' in *Adi* language	Physical barrier: To prevent rat damage in stored grains	Sarangi et al., 2009
Stored grains	Rodents	*Scorulla parasitica* : Prepare gum from the fruits, fix in sticks and put in possible rat runways for trapping	Trapping (Rao, 1996), Strong adhesive: rats get stuck to the gum	Sinha, 2010

(*Contd.*)

Crop	Insect pests	Traditional practices	Belief/Remark	Reference
Stored grains	Rodents	*Cyathula tomentosa* (Changha kakhra): Branches and inflorescence for catching rats	Spiky inflorescence	Sinha, 2010
Stored grains	Rodents	Plugging of burrows by paddy husk and local chilly land races	The suffocating pungent smokes promptly affect the respiration systems of rats and killed.	Chhetry, 2009
Stored grains	Rodents	Electric bulbs: Mix crushed electric bulbs and ground rice in equal proportion and put in possible rat runways	Rodenticide: Rats die after consuming the mixture (Corrosive)	Sinha, 2010
Stored grains	Rodents	Indigenous traps: Different types of indigenous traps made from locally available resources are put in possible rat runways for trapping rats	Trapping (Rajsekaran, 2003)	Sinha, 2010
Stored grains	Rodents	Indigenous traps of Mizoram *viz.*, *Vaithang* and *Chepthang* or *Thangchep Hnawhtawt* and *Chehrap* or *Kharkhip* or *Kawlper*	Trapping mainly based on elasticity	Azad Thakur et al., 2013
Stored grains	Rodents	Putting of cycle tyre tube in the mouth of the rodent burrows	The black cycle tube act as a rat scarer due to its snake like appearance.	Deka et al., 2006
Stored grains	Rodents	Pouring of vermilion water in the trapped rodent and releasing in the field	Due to the red colour appearance of the released rodent, the other rodent of the field runs away	Deka et al., 2006
Stored Bamboo	Rodents	Leaves of neem, *Azadirachta indica* and Karanj, *Pongamia pinnata*	Repellent	Kiruba et al., 2006

SUMMARY

- The climatic conditions of North Eastern Hill (NEH) region of India are conducive for different types of horticultural crops but at the same time these conditions are also favorable for the occurrence and multiplication of different insect pests.
- These insect pests cause colossal damage to crops resulting into low production in the region. Moreover, stored grain pests including rodents cause huge losses in every part of the world including NEH region.
- Farmers in many parts of the NEH region are still using traditional pest management practices to reduce damage by crop pests.
- Most of this traditional knowledge is published in various reports and still several things are yet to be published from different corners of the country.
- However, there is lack of comprehensive information at one place.
- Therefore, we tried to summarize all these available techniques in this chapter.

References & Recommended Reading

Anonymous, 1997. Traditional procedures and methods of storage *protection,* (GTZ, Eschborn, Germany), 8p.

Azad Thakur, N.S., Firake, D.M. and Kumar, D., 2013. Indigenous Traps for the Management of Rodent Outbreak in North Eastern Hill Region of India. *Indian Journal of Traditional Knowledge* (Accepted).

Barooah, M. amd Pathak, A., 2009. Indigenous knowledge and practices of *Thengal Kachari* women in sustainable management of *bari* system of farming. *Indian Journal of Traditional Knowledge* **8**(1): 25-40.

Bhattacharjee, P. P. and Ray, D. C., 2010. Pest management beliefs and practices of Manipuri rice farmers in Barak valley of Assam. *Indian Journal of Traditional Knowledge* **9**: 673-676.

Blackwell, W. H., 1990. Poisonous and Medicinal Plants, (Prentice Hall, New Jersey), 767.

Chanu, L. B., Chetri, G. K. N. and Sharma, G. D., 2010. Sustainable indigenous practices for the management of pest and diseases of upland rice in Manipur, North East India. *Assam University Journal of Science and Technology* **5**(1): 58-62.

Chen, H. Y., Zhou, C. X., Lou, Y. J., Duan, Z. H. and Zhao, Y., 2005. Chemical constituents from *Elsholtzia blanda. Zhongguo Zhong Yao Za Zhi*, **30**:1589p.

Chhetry, G. K. N., 2009. Indigenous pest and disease management practices in traditional farming systems in north east India: A review. *J of Plant Breeding and Crop Sci.* **1**: 28-38.

Chopra, R. N., Badhwar, R. L. and Nayar, S. L., 1941. Insecticidal and piscicidal plants of India. *J Bombay Nat Hist Soc.* **42**: 854.

Choudhury, P. A. and Leclercq, S. N., 1996. Essential oil of *Cymbopogon khasianus* (Munro ex Hack.) Bor from Northeastern India, *J Essential Oil Res.* **7**: 555.

Deka, M. K., Bhuyan, M. and Hazarika, L. K., 2006. Traditional pest management practices of Assam. *Indian Journal of Traditional Knowledge* **5**: 75-78.

Duke, J. A., Godwin, M. G. B., duCellier, J. and Duke, P. A. K., 2002. *Handbook of medicinal herbs* (CRC Press Inc, Florida, USA): 896p.

Kiruba, S., Mishra, B. P., Israel S., Jeeva S. and Dhas, S. M., 2006. Traditional pest management practices in Kanyakumari district southern peninsular India. *Indian Journal of Traditional Knowledge* **5**: 71-74.

Lalramnghinglava, J. H., 1996. Ethnobotany of Mizoram-A preliminary survey. *J Econ Tax Bot Addl Sr.* **12**: 439p.

Rao, R. R., 1981. Ethnobotany of Meghalaya: medicinal plants used by the Khasi and Garo tribes. *Econ Bot.* **35**: 4p.

Rao, R. R., 1996. Indigenous people and forests: perspectives of an ethnobotanical study from Nagaland, Northeast India. In: Ethnobiology in Human Welfare, edited by Jain, S.K., (Deep Publications, New Delhi), 367p.

Sarangi, S. K., Singh, R. and Singh, K. A., 2009. Indigenous method of rat proof grain storage by *Adi* tribes of Arunachal Pradesh. *Indian Journal of Traditional Knowledge* **8**: 230-233.

Saravanan, R., 2010. Traditional knowledge, innovation systems and democracy for sustainable agriculture: a case study on *adi* tribes of eastern Himalayas of North-east India. Innovation and sustainable development in agriculture and food, (ISDA 2010), Montpellier, June 28-30, 2010, page 1-10.

Sinha, B., 2010. An appraisal of the traditional post harvest pest management methods in northeast Indian uplands, *Indian Journal of Traditional Knowledge* **9**: 536-543.

Sinha, B., Choudhury, D. and Roy, S., 2004. Traditional practices in pest management: some examples from North-East India. In *Regional seminar on the Role of biodiversity and environmental strategies in North East India*, 1-9, online available at, http://ssrn.com/abstract=1303383.

Srivastava, S. K., Babu, N. and Pandey, H., 2009. Traditional insect bio-prospecting- As human food and medicine. *Indian Journal of Traditional Knowledge* **8**: 485-494.

Tapondjou, A. L., Miyamoto, T., Mirjolet, J. F., Guilbaud, N. and Lacaille-Dubois, M. A., 2005. Pursaethosides A-E, triterpene saponins from *Entada purseatha*, *J Nat. Prod.* **68**: 1185.

Umdor, M., 2004. Indigenous practice on protection of *Areca catechu* Linn. seedlings: A case study in Meghalaya. *Indian Journal of Traditional Knowledge* **3**: 253-255.

Chapter – 6

Indigenous Pest Management Practices in North East India - An Overview

Arunava Samanta, Archan Kanti Das and Sandip Patra

INTRODUCTION

North East India is a rich storehouse of indigenous agricultural practices and local varieties of crops. The region comprising of the states of Arunachal Pradesh, Assam, Meghalaya, Manipur, Tripura, Mizoram, Nagaland and Sikkim can be physiographically categorized into the Eastern Himalayas, Northeast hills (Patkai-Naga Hills and Lushai Hills) and the Brahmaputra and Barak Valley plains. At the confluence of the Indo-Malayan, Indo-Chinese and Indian biogeographically realms, the NE region is unique in providing a profusion of habitats, which features diverse biota with a high level of endemism (Chatterjee et al., 2006). The northeastern region is rich in biological diversity and contains more than one-third of the country's total biodiversity. This biodiversity and variability of the ecosystem are used and conserved by traditional communities through various informal institutions and using traditional ecological knowledge systems. Ethnic communities of the region practice their traditional cultivation practices of diverse crop suitable in different ecosystems solely on organic conditions without any external inputs to the system; and value the land and the environment for the sustainability of the system for future generations. There is no systematic record to describe what they are, what they do and how they do, how they can be changed, their operations, their boundaries and their applications.

Most of the indigenous knowledge disappears due to the interference of foreign technologies and development concepts that promise short-term gains or solutions to problems without being capable of sustaining them. The beginning of the concept of sustainable agriculture in late eighties in Indian agricultural scenario has evoked significance on indigenous technical knowledge (ITK) that has the element of use of natural products to solve the various problems of agriculture and allied activities. The improvement of the quality of life of the Indians who in great majority live in and depend on agricultural production systems would be impossible by keeping this rich tradition of ITK aside. In this context, the purpose of this chapter is to collect the scattering information about traditional pest management practices in North Eastern states and compiled together to make a concrete document especially on indigenous insect pest management of this region.

INDIGENOUS TECHNOLOGICAL KNOWLEDGE (ITK)

Traditional farming system is an ecologically based age-old farming system developed by ancient farmers through generations of their interaction with nature and natural resources for food, fodder and fibre. Indigenous knowledge is the knowledge of the indigenous people inhabiting different geographical regions of the world with their own language, culture, tradition, belief, folklore, rites and rituals. The increasing attention on indigenous knowledge is receiving by academia and the development institution has not yet led to a unanimous perception of the concept of indigenous knowledge. None of the definitions is essentially contradictory; they overlap in many aspects.

Warren (1991) and Flavier (1995) present typical definitions by suggesting: Indigenous knowledge (IK) is the local knowledge – knowledge that is unique to a given culture or society. IK contrasts with the international knowledge system generated by universities, research institutions and private firms. It is the basis for local-level decision making in agriculture, health care, food preparation, education, natural-resource management, and a host of other activities in rural communities.

According to Haverkort (1995), Indigenous Traditional Knowledge is the actual knowledge of a given population that reflects the experiences based on tradition and includes more recent experiences with modern technologies.

IMPORTANCE OF INDIGENOUS KNOWLEDGE

In the emerging global knowledge economy a country's ability to build and mobilize knowledge capital, is equally essential for sustainable development as the availability of physical and financial capital (World Bank, 1997). Indigenous knowledge is the basic component of any country's knowledge system. It is the combination of skills and experiences to improve their livelihood. Indigenous knowledge have major role to develop global knowledge, for instance in medicine and veterinary medicine with their intimate understanding of their environments. It is practices continuously from generation to generation and modifying time to time with the changing environment. Indigenous knowledge is also the social capital of the poor, their main asset to invest in the struggle for survival, to produce food, to provide for shelter or to achieve control of their own lives. Indigenous knowledge can be utilized in different sectors viz. agriculture, animal husbandry and ethnic veterinary medicine, use and management of natural resources, primary health care (PHC), preventive medicine and psychosocial care, saving and lending, community development thereby enhances developmental process.

INDIGENOUS PEST MANAGEMENT PRACTICES FOR DIFFERENT CROPS IN NORTH EAST INDIA

Cereals

Rice is the major food grain crop of the North Eastern state followed by maize, occupying 3.51 million hectares which accounts for more than 80% of the total cultivated area of the region. Several insect pests attack on rice in both upland and low land conditions of the entire north eastern region. Traditional land races cultivars of rice varieties which are in practice since generations in the region by diverse ethnic groups are successfully managed against pest pathogens and nutrient deficiencies using their indigenous/ traditional knowledge in location specific areas. They manage insect pests in rice ecosystem with their locally available resources since age-old. Some indigenous pest management practices in rice are presented in Table 1.

Fruits and Vegetables

North East region of India is one of the richest reservoirs of genetic variability and diversity with different kinds of fruits and vegetables. Among fruit crops, citrus, mango, banana and some others fruits and all type of vegetables are grown in this region. Due to unique climatic condition all fruit and vegetable crops are attacked by different pests and causing considerable damage to the crops. Farmer of this region protects their crops from pest problem by adopting their traditional practices. Few indigenous pest management practices for fruit and vegetable crops are documented which are presented in Table 2.

Table 1: Indigenous management of rice pests in north east India

S. No.	Practices	Target pests	Rationale	Place	References
1.	Mixing of cow dung with water in the rice field	Rice hispa *(Dicladispa armigera)*	Cow dung contains Nitrogen, helps to revive the damage caused by rice hispa	Assam	Deka et al., 2006
2.	Placing of *Nishinda* plant twigs (*Vitex negundo* L.) in the infected rice fields	Rice hispa *(Dicladispa armigera)*	The foul smell of the plant repels the pest	Barak Valley, Assam	Bhattacharjee and Ray, 2010
3.	Rearing duck near the paddy field	Rice hispa (*Dicladispaarmigera*)	Ducks voraciously feed on adult rice hispa	Assam	Deka et al., 2006
4.	Smashed crabs/eels are putted on the top of a bamboo stick and placed in the crop field	Gundhi bug (*Leptocorisa variconis* and *L. acuta)*	Rotten crabs /eels attract gundhi bug	Meghalaya and Assam	Rautray, 2002 and Sinha et al., 2007
5.	Dead frogs/crabs are placed on the top of the stick. Sometimes farmers also seen to tie pieces of salted fish with each stick	Gundhi bug (*Leptocorisa variconis* and *L. acuta*)	As the dead frogs/crabs get rotten and start to emit foul smell, adult gundhi bugs are attracted by this foul smell and gather in mass on the rotten frogs/crabs	Arunachal Pradesh	Rautray, 2002
6.	Keeping of grape fruit (*Citus paradisica*) in the rice field at flowering or milky stage	Gundhi bug, (*Leptocorisa variconis* and *L. acuta*)	Gundhi bugs are repelled away with the smell and odour emending from the peels of grape fruit	Assam	Rautray, 2002
7.	Burning of bicycle tyres near the rice field before panicle initiation	*Gundhi*bug (*Leptocorisa acuta*)	The foul smell of burned tyres repel *Gundhi* bug	Assam	Deka et al., 2006

(*Contd.*)

S. No.	Practices	Target pests	Rationale	Place	References
8.	Burning of clothes on the four corners of the rice fields	*Leptocoryza* sp.	Attracted to the fire and dies	Barak Valley, Assam	Bhattacharjee and Ray, 2010
9.	Placing of twigs of *Clerodendrum oryzae* L. randomly in the infected rice fields	*Leptocoryza* sp.	Kuthap leaves repel the insect	Barak Valley, Assam	Bhattacharjee and Ray, 2010
10.	Cut the bamboo shoots into pieces and spray in rice fields after soaking in water overnight	Leaf roller	-	Arunachal Pradesh	Sarma et al., 2002
11.	Removal of grassy weeds from bunds around the paddy field	Leaf folder (*Cnaphalocrocis medinalis*)	The grassy weeds may provide suitable microhabitats.	Assam	Deka et al., 2006
12.	Well fermented wine pomace (Wine residue) usually made up of millets are placed at the source of irrigation canal of terrace rice fields	Leaf folder	Unpleasant odour of pomace repels the insect	NorthEast India	Chhetry and Belbahri, 2009
13.	Erection of bamboo (*Bambusa indica*) branches or top or other stick in the rice field	Stem borer (*Scirpophagaincertulus*) andSwarming caterpillars	Attract birds to take rest on the bamboo stick. The birds feed the adult moths of stem borer and swarming caterpillars	Assam	Anonymous, 1948
14.	Steam decoction of neem (*Azadirachta indica*	Stem borer (*Scirpophaga*	The azadirachtin present in the neem seed and leaf act as	Assam	Bhagavati, 2001

(*Contd.*)

S. No.	Practices	Target pests	Rationale	Place	References
	A. Juss.) seeds and leaves, and the extract is sprinkled in the rice crop. The extract is prepared by mixing 1 to 3 gm of ground neem seed/leaf in 1 l of water for 12 hrs	*incertulus)*	antifeedant and growth retardant to insect		
15.	Keeping slices of pumalo (*Citrus grandis* Osbeck) in the paddy field @1 trap/$6m^2$	Stem borer, *Scirpophaga incertulus*	The essential oils of pumalo repel stem borer (Pathak, et al., 2001)	Assam	Deka et al., 2006
16.	Placing of chopped leaves of Indian rhododendron or Phutuka (*Melastoma malabathricum* Linn.) in paddy field	Stem borer, *Scirpophaga incertulus*	The chopped leaves emit foul smell, which drive away the insect from field	Assam	Deka et al., 2006
17.	Placing of grounded bark of drumstick (*Moringa oleifera* Lam.) in the field	Stem borer, *Scirpophaga incertulus*	Bark may contain insecticidal principles. Drumstick has been reported to have medicinal Properties (Vaidyaratnam, 1995)	Assam	Deka et al., 2006
18.	Placing few branches of fern (*Cybotium* sp.)	Stem borer, *Scirpophaga incertulus*	Fern has insecticidal activity (Bhuyan, 2003)	Assam	Deka et al., 2006
19.	Burying the puthi or barb fish (*Puntius* sp.) in the soil for 15-20 days and spraying water	Stem borer, *Scirpo phaga incertulus*	The fermented fish extract repels stem borer	Assam	Deka et al., 2006

(Contd.)

S. No.	Practices	Target pests	Rationale	Place	References
	extract in the rice field				
20.	Placing citrus or *sakala* tenga (*Citrussinensis* Osbeck) peels in the rice field @ 1 trap/6m^2	Stem borer, *Scirpophaga incertulus*	The citrus peels have insect repellent properties (Pathak, et al., 2001; Anonymous, 1950)	Assam	Deka et al., 2006
21.	Rope, dipped in kerosene stirred in the water of rice field	Rice case worm (*Nymphula depunctalis*)	Kerosene act as toxicant to case worm (Srivastava, 1993)	Assam	Deka et al., 2006
22.	Spreading of raw cow dung in the field water	Rice case worm (*Nymphula depunctalis*)	Water in the rice field mixed with cow dung has oxygen deficiency	Assam	Deka et al., 2006
23.	Raw cow dung is mixed with water to prepare a suspended solution, which is sprayed in the rice field	Thrips (*Thrips oryzae*)	Odour of the cow dung repels the pests	Assam	Deka et al., 2006
24.	Oak tree bark are grounded and placed over the source of irrigation canal	Brown Plant Hopper (BPH)	-	Northeast India	Chhetry and Belbahri, 2009
25.	Spreading of goat dung in the rice field	Different insect pests	Smell of goat dung may repel insects	Assam	Deka et al., 2006
26.	Putting of cycle tyre tube in the mouth of the rodent burrow	Rodent	The black cycle tube act as a rat scarer due to its snake like appearance	Assam	Deka et al., 2006
27.	Pouring of vermilion water in the trapped rodent and releasing in the field	Rodent	Due to the red colour appearance of the released rodent, the other rodent of the field runs away	Assam	Deka et al., 2006

Table 2: Indigenous pest management in fruit and vegetable crops

Sl. No.	Practices	Target pests	Rationale	Place
Vegetable				
1.	Smoking in the pumpkin field	Fruit fly *Bactocera cucurbitae*	Smoke often act as repellent to various insect-pests	Assam
Citrus				
1.	Application of fish water on the base of citrus plant	Citrus trunk borer (*Anoplophora verstegii*)	The fish water attract predatory red tree ant (*Oecophyla* sp.)	Assam
2.	Smoking below the citrus plant at the time of flowering	Citrus insect pests	Smoke repels insect pests	Assam
3.	Placing of red tree ant (*Oecophylla smoragdina*) nest on the citrus plant	Citrus insect pests	Red tree ant is a predator of citrus pests	Assam
Coconut				
1.	Placing of long hair of women in the crown portion of coconut tree	Rhinoceros beetle, (*Oryctes rhinoceros*)	The beetle engulfed with long hair due to which they cannot move	Assam
2.	Placing of a dead frog at the base of the coconut plant @ 1 trap/m^2	*Rhinoceros beetle, Oryctes rhinoceros*.	The dead frog repels insect (Pathak et al., 2001)	Assam
3.	The fragmented human hair and dry fish mixture are kept in the crown	Squirrel	Along with the fish the fragmented hairs are also consumed, which produces adverse effects to the squirrel	Assam

Source: Deka et al., 2006.

Stored products

The post harvest losses to food grains and seeds in the north-eastern region of India is much higher than other parts of the country as the climatic conditions prevailing in the region is very much conducive for rapid growth and multiplication of stored product insects. Major part of the food grain produced is stored at the farm level to ensure household food supply. People store food grain for future cash need and also for seed purpose. Thus, post harvest protection of grains and seeds assume a greater importance in over all crop protection system. In this region, there are about a dozen species of insect pests which are attacked to the food grain in storage. To protect stored food grains from insect and non-insect pests, tribal people of this region mainly used plant parts which are easily available in their locality. Some post harvest pest management practices for this region are presented in Table 3.

Table 3: Indigenous post- harvest pest management

Sl. No.	Practices	Target pests	Rationale	Tribe/Place	References
1.	Put fresh or dried leaves/branches of *Artemisia vulgaris* L. in and around granaries	Stored insects and rats	Aroma of the shrub is intolerable	*Tangkhul*, Ukhrul, *Mao*, *Poumai*, Senapati	Sinha, 2010
2.	Crush and put fresh or dried leaves of *Azadirachta indica* Juss. and *Cannabis sativa* L. in and around the granaries	Stored insects	Repellent	*Tangkhul*, Ukhrul	Sinha, 2010
3.	Branches of *Cyathula tomentosa* Moq. with inflorescences are put in and around granaries	Rats	Inflorescence stick to rat's body	*Mao, Poumai*, Senapati	Sinha, 2010
4.	Crush and put fresh or dried leaves of *Cymbopogon khasianus* Stapf ex Bor in and around granaries	Stored insects	Repellent	*Poumai*, Senapati, *Tangkhul*, Ukhrul	Sinha, 2010
5.	Place fresh leaves of *Dendrocnide sinuate* (Bl.) Chew in between two layers of boiled rice, after 24 hrs., put treated rice in possible runways as rodenticides	Rodents	Leaves contain toxic substances	Garo	Sinha, 2010
6.	Crush and put fresh or dried leaves of *Elsholtzia blanda* Benth in and around granaries	Stored insects	Repellent	*Mao*, Senapati	Sinha, 2010
7.	Mix equal proportion of ground seed kernel of *Entada pursaetha* DC and ground rice and put in possible rat runways as rodenticide	Rodents	Seeds are toxic to rats	*Poumai*, Senapati, Garo	Sinha, 2010

(Contd.)

Sl. No.	Practices	Target pests	Rationale	Tribe/Place	References
8.	Crush and put fresh or dried leaves of *Melia composita* Wild in and around granaries	Stored insects	Leaves are repellent to insect pests	*Poumai*, Senapati	Sinha, 2010
9.	Rub pulses with pork oil before storing	Stored insects	Protect pulses from post harvest damage	*Tangkhul*, Ukhrul	Sinha, 2010
10.	Keeping neem leaves and branches over the stored rice	Rice weevil (*Sitophilus oryzae*)	Neem leaves repel the insects	Barak Valley, Assam	Bhattacharjee and Ray, 2010
11.	Keeping Heigri (*Dillenia indica* L.) leaves over the stored rice	Rice weevil (*Sitophilus oryzae*)	Leaves repel the weevil	Barak Valley, Assam Ray, 2010	Bhattacharjee and
12.	Keeping of *neem* (*Azadirachta indica*) leaf, or *ber* (*Ziziphus jujuba* Mill.) branch over the stored rice	Control rice weevil (*Sitophilus oryzae*)	The neem and ber leaf repel the weevil	Assam	Deka et al., 2006
13.	Keeping of Chawai sabi (*Polygonum hydropiper* Linn.) leaves over the stored rice	Rice moth (*Sitotroga cerealella*)	Leaves repel the moth	Barak Valley, Assam	Bhattacharjee and Ray,2010
14.	Keeping a pot of water over the stored rice grain	Rice moth (*Sitotroga cerealella*)	The moth falls on the water surface and thus killed	Assam	Deka et al., 2006
15.	Placing the Tezpata (*Cinnamomum tamala* Nees & Eberm) leaves over the stored grain	Stored grain pests	Odour of Tezpata repels many stored grain pests	Barak Valley, Assam	Bhattacharjee and Ray, 2010
16.	Placing of curry leaves (*Murraya koenigii* Spreng.) in grain storage	Stored grain pests	Odour of curry leaves repels many stored grain pests	Assam	Deka et al., 2006
17.	Mixing of mustard oil with stored grains	Stored grain pests	The mustard oil repels stored grain pests	Assam	Deka et al., 2006
18.	Sand cover (2-5 cm depth) over the stored potato	Potato tuber moth (*Pthorimaea operculella*)	The female moth can not lay egg on the potato	Assam	Deka et al., 2006

Spices

Rahman et al. (2009), reported that the farmers in this region have been following various indigenous practices for management of major insect pests in ginger like white grub (*Holotrichia*) and shoot borer (*Dichocrocis punctiferalis*). The adult beetles of white grub (*Kumlay kheera* in Nepali) after emergence settle on trees like *Ficus hookerii*. Farmers collect and destroy them. The use of unrotton cow dung is avoided in white grub prone areas as it carries the eggs.

OTHER INDIGENOUS PEST MANAGEMENT PRACTICES

Chhetry and Belbahri (2009) reported some other ITK practiced by farmers for pest management of NE India are:

a. Farmers maintain good seed health from attack of pests for sowing in the next season through:

    i. Hanging over fire furnace/ kitchen for constant smoking

    ii. Mixing with ashes of fire wood

    iii. Smoking well dried healthy seeds with edible and non- edible oils

    iv. Mixing with neem seed powder and storing the same in a seed bin.

    v. Using aromatic plants such as citronella grass, lemon grass, peels of pomelo against maize weevil by mixing/placing these plants over maize grains granary.

b. Seeds of maize and leguminous crops are often kept intact along with their outer husk and hang over the kitchen/furnace. Here, maize cobs are tied up in bunches of 10-12 cobs by folding their next to outermost husk and hang over the wooden beams of kitchen and sometimes roof beam in the periphery of the house.

c. To protect the granary from pest and pathogens they specially prepare bamboo granaries plastered with mixture of fresh cow dung and mustard oil cakes for storing paddy, on the top of which branches with leaves of *Zanthoxylum acanthopodium* are placed to keep away pest, mostly white butterflies, a common pest of the stored paddy grains. Paddy granaries are either placed near the kitchen or vicinity to kitchen in a separate house.

d. Foliar application of wood ashes in the wee hours of the day keeps away aphids, pod borers and diseases from plants especially in vegetables.

e. Dusting of finely ground tobacco leaves keep the aphid pest and diseases away from plants.

f. Hookah water is very much effective for controlling pest and diseases of major and minor crops such as blast of rice, pod borers, sucking bugs of vegetables etc.

g. Fish and meat wash water application is also a mild deterrent in keeping away of pest due to unpleasant environment for the proliferation of pest and pathogens.

h. Nematodes of turmeric, tomato, chillies and ginger are controlled by either intercropping with *Chrysanthemum coronarium, Tagetes erecta,* or growing *Tagetes erecta* as border crops. This is a very effective method and often farmers incorporate leaves of these trap crops into the soil to enhance effectiveness and nutrients enrichment of crops.

i. Burning of slashed debris which kills the resident pests and pathogens from the system because many plants and grasses serve as the alternate host of crops. This clean cultivation practices enable farmers to harvest crops less infected by pest and pathogen.

j. Mixed cultivation of rice with sparsely grown maize, legume crops, job's tear (*Coix lacryma jobi. L)*, shorghum and ground vegetables, protects from diseases and pest of rice probably due to the physical barriers of intercrops in the movement of air borne propagules, augmenting microclimate and humidity etc. Maize and sorghum not only provide food but also acts as perch for birds to feed on insects and pest of paddy in jhum field.

k. Keeping tree boles / trunk and partially cut older trees amidst jhum field which facilitates perching of diverse birds for selective prey of pest such as brown grasshopper of paddy, pod borers of pigeon pea etc.

l. Elaborate preparation of land into ridges for soil conservation and furrows for channelling water which one often finds in high and mid elevation *jhum* land where fallen *pinus* needles, twigs and slashed bushes are burnt by covering with a thin layer of soil drag from furrows. In this technique biological material and soil get burnt slowly by suffocating the whole ridges with smoke. Thus, the whole system becomes sterile making the soil free from soil based pest and pathogen and therefore crops such as potato and ginger raised by employing this technique are often healthy.

m. Carnivorous birds such as Egrets and others which prey on leaf folder larvae. Stem borer, grass hoppers etc. of rice in the sedentary wet land rice are facilitated by erecting suitable perches in sufficient numbers. Sacred grooves conserved by community also facilitate perching of birds which effectively manage the unwanted pest from rice field.

n. Farmers often plough their field repeatedly soon after harvest for exposing soil inhabiting insect pests, arthropods, nematode etc. to harsh weather and to facilitate natural predators. Insects such as grasshopper, crickets and borers lay their eggs in the upper layer of soil in paddy fields eventually exposed during the course of repeated operation and their eggs either desiccate or preyed by the Egrets and their natural predators. This indigenous repeated ploughing technique for getting rid of soil borne insects which damage paddy crops is very effective in wet land system and dry sedentary farming system.

o. To control rodents, burrows of rats are stuffed with smoke by burning paddy husk and land race dry chillies variety. Complete control over rodents depends on the number of burrows plugged by smokes.

SUMMARY

- North East India is a rich storehouse of indigenous agricultural practices and local varieties of crops.
- Most of the indigenous knowledge's are disappearing due to the interference of foreign technologies and development concepts that promise short-term gains or solutions to problems without being capable of sustaining them.
- The beginning of the concept of sustainable agriculture in late eighties in Indian agricultural scenario has evoked significance on indigenous technical knowledge (ITK) that has the element of use of natural products to solve the various problems of agriculture and allied activities.
- Indigenous knowledge can be utilized in different sectors *viz.* agriculture, animal husbandry and ethnic veterinary medicine, use and management of natural resources, primary health care (PHC), preventive medicine and psychosocial care, saving and lending, community development thereby enhances developmental process.
- Indigenous pest management knowledge derived through long experiences and perceptions accumulated by traditional farmers during the course of their interactions with the nature and natural resources need to be effectively integrated and should not be viewed in isolation.

- Indigenous pest management yet needs thorough validation in view of changing agricultural scenario from traditional to integrated farming system through inorganic and organic farming methods.
- Some of this validated indigenous traditional knowledge (ITK) may be incorporated as an integral component into the IPM concept for evolving better pest management strategies.
- All the academics, policy makers, planers should pay greater attention to this invaluable wealth of knowledge that is threatened by extinction. Through systematic approach it is necessary to protect this fast disappearing ITK under the influence of modern agriculture as well as preserve the indigenous pest management identity of farming communities of this country.
- Therefore, this rich heritage of the county should be harnesses, preserved and documented in the interest of future generations and developed as indigenous integrated pest management before they are lost.

CONCLUSION

Formal pest management knowledge derived through hypothetic-deductive method whereas indigenous pest management knowledge derived through long experiences and perceptions accumulated by traditional farmers during the course of their interactions with the nature and natural resources need to be effectively integrated and should not be viewed in isolation. Indigenous pest management yet needs thorough validation in view of changing agricultural scenario from traditional to integrated farming system through inorganic and organic farming methods. Some of this validated Indigenous traditional knowledge (ITK) may be incorporated as an integral component into the IPM concept for evolving better pest management strategies. It is very clear from the discussion that there is much to be learned from the indigenous knowledge systems of local people. All the academics, policy makers, planers should pay greater attention to this invaluable wealth of knowledge that is threatened by extinction. Through systematic approach we have to protect this fast disappearing ITK under the influence of modern agriculture as well as preserve the indigenous pest management identity of farming communities of this country. Therefore, this rich heritage of the county should be harnesses, preserved and documented in the interest of future generations and developed as indigenous integrated pest management before they are lost.

References & Recommended Readings

Anonymous, 1948. The wealth of India, Raw Materials, Vol 1, (Council of Scientific & Industrial Research, New Delhi), 254.

Anonymous, 1950. The wealth of India, Raw Materials, Vol II, (Council of Scientific Industrial Research, New Delhi), 427.

Bhagavati, R. R., 2001. Homemade agricultural medicine (Assamese). *Mrittika* **3**(2): 43-45.

Bhattacharjee, P. P. and Ray, D. C., 2010. Pest management belief and practices of Manipuri rice farmers in Barak Valley, Assam, *Indian Journal of Traditional Knowledge* **9**(4): 673-676.

Bhuyan, M., 2003. Studies on some potential insect control agents from plants of North East India, PhD Thesis, Dibrugarh University, Dibrugarh, Assam.

Chatterjee, S., Saikia, A., Dutta, P., Ghosh, D., Pangging, G. and Goswami, A. K., 2006. Background paper on biodiversity significance of North East India for the study on natural resources, water and environment nexus for Development and Growth in North Eastern India, WWF-India, 172 B Lodi Estate, New Delhi 110003.

Chhetry, G. K. N. and Belbahri, L., 2009. Indigenous pest and disease management practices in traditional farming systems in north east India-A review. *Journal of Plant Breeding and Crop Science* **1**(3): 28-38.

Deka, M. K., Bhuyan, M. and Hazarika, L. K., 2006. Traditional pest management practices of Assam, *Indian Journal of Traditional Knowledge* **5**(1): 75-78.

Flavier, J. M., 1995. The regional program for the promotion of indigenous knowledge in Asia, pp. 479-487 in Warren, D. M., Slikkerveer, L. J. and Brokensha, D. (eds). *The cultural dimension of development: Indigenous knowledge systems*. London: Intermediate Technology Publications.

Haverkort, B., 1995. Agricultural Development with a Focus on Local Resources: ILEIA'S view on Indigenous Knowledge. In The cultural dimensions of development: Indigenous Knowledge Systems (Eds. D. M. Warren, Slikkerveer, L. J. and Brokensha, D.). Intermediate Technology Publications Ltd., London, 454 - 457.

Pathak, K. A., Thakur, N. S. A., Rao, K. R. and Shylesha, A. N., 2001. Insect pests of crops and their management, in: Steps Towards Modernization of Agriculture in NEH Region, ed Verma, N. D. and Bhatt, B. P. (Indian Council of Agricultural Research, New Delhi), 121-159.

Rahman, H., Karuppaiyan, R., Kishore, K. and Denzongpa, R., 2009. Traditional practices of ginger cultivation in Northeast India, *Indian Journal of Traditional Knowledge*, **8**(1): 23-28.

Rautray, S. K., 2002. Indigenous knowledge and belief on rice farming in eastern India. *Indian J. Hill Farming* **15**: 44-49.

Sarma, B. K., Singh, J. K. and Annadurai, A., 2002. Collection of multi crop diversity from the eastern Himalayan region. *Indian J. Hill Farming* **15**: 94-99.

Sinha, B., Singha, R. and Choudhury, D., 2007. Ecological pest management for emerging pest problems. Leisa India December: 8-9.

Srivastava, K. P., 1993. A Text Book of Applied Entomology, Vol II, (Kalyani Publishers, Ludhiana), 77p.

Vaidyaratnam, P. S., 1995. Varier's Arya Vaidya Sala, (Orient Longman Ltd). *Indian Medicinal Plants* **4**.

Warren, D. M., 1991. The role of Indigenous Knowledge in facilitating the agricultural extension process". Paper presented at International Workshop on Agricultural Knowledge Systems and the Role of Extension. Bad Boll, Germany.

World Bank, 1997. Knowledge and Skills for the Information Age, The First Meeting of the Mediterranean Development Forum; Mediterranean Development Forum. URL:http://www.worldbank.org/html/fpd/technet/mdf/objectiv.htm

Chapter – 7

Notes on the Ethno-Veterinary Practices Used for Treatment of Different Ailments in Tripura State, India

Partha Pratim Bhattacharjee, J. Majumder, R. Lodh and B. K. Agarwala

INTRODUCTION

Ethno-Veterinary Medicine (EVM) or Veterinary Anthropology refers to holistic and interdisciplinary study of traditional knowledge, skills, methods, practices and folk beliefs of the people about the health care of animal husbandry and production of livestock (McCorkle, 1986). It encompasses information on diseases and their control; remedies and clinical practices for treatment and prevention; management, feeding and breeding strategies; spiritual elements; and the use of human resources that hold the information and experience (Mathias, 2004).

India has a rich heritage of traditional health care system for livestock through Ayurvedic, Unani, and Homoeopathy system of medicines which are used since time immemorial. Our knowledge in ethno-medicine is available through Vedic texts and commentaries (Priyadarsan, 1991). There is another less exploited source of information in use which comes from folklore passed on through generations in remote habitations (Jain, 1966).

For most of the tribal communities there is little published data on the ethno-Veterinary plants used by them despite the fact that medicines made from herbs has been very crucial for the animal healthcare of most developing countries. In absence of proper documentation, much effort is needed in research and integration of existing knowledge. On the other hand, plants of ethno-veterinary importance and traditional knowledge of ethno-veterinary medicines are gradually depleting due to lack of proper records as well as over exploitation (Jain, 1996).

Tripura is a small hilly state of north-eastern India with a rich diversity and density of plants. Out of the total area of 10,492 km^2 of the state, forest covers an area of about 6,292.681 km^2 with records of 379 species of trees, 320 of shrubs, 581 of herbs, 165 of climbers, 16 of climbing shrubs, 35 of ferns and 45 of epiphytes. Tripura has a maximum plant diversity index of 5.23 in India (Kshirsagar and Upadhyay, 2009).

In many parts of India, plants are used for treatment of domestic animals, and many studies have been done in this regard, *viz*., in Sikkim (Bharati and Sharma, 2009); in central Himalaya (Iswar, 1981, Bisht et al., 2004, Pandey and Joshi, 1999, Gaur et al., 1992, Samal et al., 2002, 2003, Tiwari and Pande, 2004, Tiwari et al., 2007); in central India (Mishra et al., 2010); in Uttar Pradesh (Nigam and Sharma, 2010, Kumar et al., 2012); in Kerala (Thomas et al., 2011); in Karnataka (Harsha et al., 2005); in Tamil Nadu (Kiruba et al., 2006); in Uttarakhand (Tiwari and Pande, 2010); in Bihar (Vidyanath et al., 1991); in Orissa (Satapathy, 2010, Mishra, 2011); in Maharastra (Salave et al., 2011, 2012); in Andhra Pradesh (Reddy and Sudarsanam, 1987); in Rajasthan (Sebastine, 1984); in Assam (Borthakur and Sharma, 1994) and in West Bengal (Bandyopadhyay and Mukherjee, 2005, Pandit, 2010). These included plants used as veterinary medicine by different tribal communities of India (Maheswari et al., 1986, Sebastine and Bhandari, 1984). However, no such account was available from the Tripura state.

In view of the above, the present study was made to document the information of indigenous veterinary medicinal plants and traditional knowledge used by the local people of Tripura for their domestic livestock.

MATERIALS AND METHODS

Study area

Tripura is situated in the north-eastern part of India and shares international boundary of 837 km with Bangladesh on north, west and south, and on the east it share boundary of 112.7 km with the states of Assam and Mizoram. It is one of the smallest states in India with a total area of 10,492 km^2 and lies between 22°56' to 24°32' N latitude and 90°09' to 92°20' E longitude (Fig. 1).

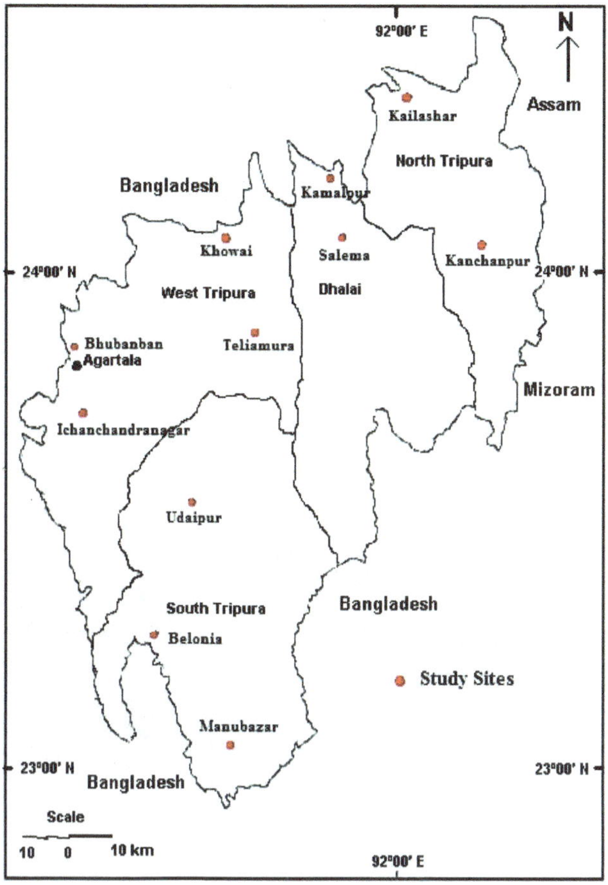

Fig. 1: Map of Tripura state showing different study sites

The climate of the state is generally moist and humid. The minimum and maximum temperatures in summer are 21°C and 38°C and, in winter, it ranges between 4°C and 33°C. The average annual rainfall varies between 2250 mm to 2500 mm. The altitude varies from 15 m to 850 m. Major forest types occurring in the state are tropical moist deciduous and sub-tropical broad leaf forests. Bamboos are predominant in the forests of the state. Heterogeneous topography of the state facilitates rich assemblage of floral and faunal biodiversity. However, the increasing influences of anthropogenic factors have created a large number of edges in the forests and are a cause in deterioration of natural ecosystem.

Ethno-veterinary plant survey

Participatory methods were used for the collection of information. Eleven sites were selected for the study (Fig. 2 and Table 1). All the sites were visited in different seasons to avail the plant resources in their flowering condition during the period from September 2009 to August 2011. The field work comprised two approaches; first, the survey part consisted of visit to individual site where study was conducted with the cooperation of local healers. A semi-structured questionnaire was prepared in local language and interviews were taken from them. The information for particular ailments was cross checked through the repeated queries during the interviews to get accurate and valid information. This was followed by visits to the actual sites of plants for their collection and identification (Plate I). The collected specimens were dried and kept as herbarium and part of those were prepared in to crude drugs. Secondly, an inventory of the method of preparation of crude drugs, local names of plants, their parts used, and method of application as medicine, against various ailments was made.

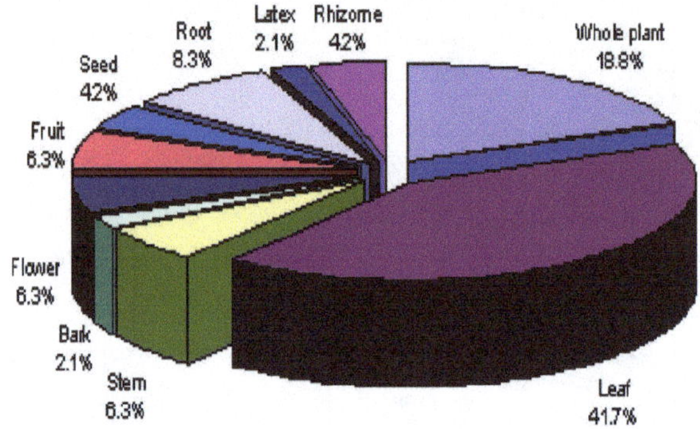

Fig. 2: Utilization of plant parts of different medicinal plant species in Tripura

Table 1: Geo-coordinate details of the study sites

Sl.No.	Sites	Coordinates	Altitude (m)
1.	Bhubanban	23°51'56.50"N; 91°1573.70"E	21
2.	Ichanchandranagar	23°45'31.02" N;91°15'02.41"E	24
3.	Khowai	24°04'07.20" N; 91°36'50"E	40
4.	Teliamuna	23°51'37.30" N; 91°37'28.76E	51
5.	Kamalpur	24°07'54.32""N;91°49'39.47"E	52
6.	Salema	24°00'44.79" N; 91°50'02.22"E	75
7.	Kailasahar	24°18'25.61" N;92°00'40.87"E	32
8.	Kanchanpur	24°01'52.17" N;92°12'04.38"E	68
9.	Udaipur	23°32'08.10" N;91°28'43.35"E	28
10.	Belonia	23°14'53.73" N; 91°28'14.03"E	25
11.	Manubazar	23°04'36.50" N; 91°38'51.39"E	24

Acalypha indica L.

Amaranthus spinosus L.

Curcuma domestica Val.

Eclipta alba (L.) Hassk.

Streblus asper Lour

Ocimum sanctum L.

Terminallia bellirica (Gaertn) Roxb.

Vernonia cinera (L.) Less

Annona squamosa L.

Ficus hispida L.f.

Plate I

RESULTS AND DISCUSSION

A total of 45 plant species under 30 families and 45 genera were documented which are used by the local people as ethno-veterinary medicines. The information on scientific names, families, local names of the plants, part(s) used to cure ailments and methods of dosage are provided in Table 2.

Among the different plant parts, leaves are found to be used more frequently (41.7%) for medicines followed by roots (8.30%), flowers/fruits, stems and whole plants (18.8%) (Fig. 2) for different ailments of domestic animals (Table 2). Nine plant species *viz.*, *Amaranthus spinosus* L., *Androdraphis paniculata* (Burm.f.) ex Nees, *Cynodon dactylon* (L.) Pers., *Echinochola crusgalli* (L.) P. Beauv., *Leucas aspera* (Willd.) Link, *Polygonum hydropiper* L., *Solanum xanthocarpum* Schrad. & Wendl., *Spermacoce hispida* L. and *Tagetes erecta* L. are used as whole plant for administration of medicine (Table 2). The recorded ethno-botanical plant species included herbs (42.22%), trees (26.67%), shrubs (13.33%), climbers (8.89%), and grasses (8.89%) (Fig. 3). These plants are used to treat constipation, wound recovery, cough and cold, external parasites, dysentery, worm infection, skin diseases, arthritis, dog bite, bone fracture, fever, loss of appetite, tongue infection, bronchial problems etc, and also for health improvement and for increasing lactation.

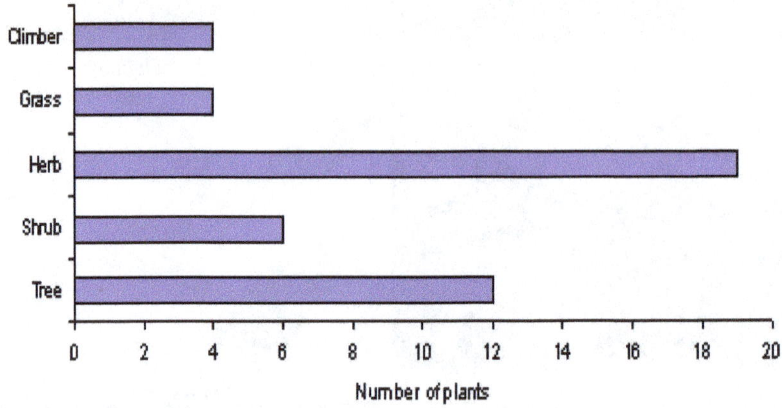

Fig. 3: Use pattern of plant types and their number of species recoreded for the ethno-veterinary medicine value in Tripura

Most of the people using the plant species to treat the animals got this knowledge from their forefathers, with exception of a few, who learned it from others or by hit and trial method. It is important to record here that most of the people who participated in this study were smallholders and could not afford modern drugs, so it was easy for them to get plants from local areas and to treat their livestock with home made preparation.

Table 2: List of ethno-veterinary medicinal plants used in Tripura for treatment of different aliments of domestic animals

S.No.	Botanical name	Local name	Family	Plant type	Parts used	Ethno-veterinary practices
1.	*Acalypha indica* L	Muktajhuri	Euphorbiacae	Herb	Leaf	10-15 gm of leaf past is given orally for constipation problem, leaf paste is also used externally for wound recovery
2.	*Adhatoda vasica* Nees	Basak	Acarthaceae	Shrub	Leaf	50 gm leaf paste is given daily for a week for healing from cough and cold
3.	*Aegle mamelos* (L.) Corr.	Bel	Rutaceae	Tree	Leaf	Leaf paste is used for external parasites
4.	*Alstonia scholaris* (L.) R Br.	Chatim	Apocynaceae	Tree	Latex	1-2 ml latex is given orally in dysentery, also extremely applied in wounds
5.	*Amaranthus spinosus* L	Katanate	Amaranthaceae	Herb	Whole plant	Whole plant is boiled and given to cattle for increasing lactation
6.	*Androdraphis paniculata* (Burm.f.) ex Nees	Kalamegh	Acantheceae	Herb	Whole plant	10 gm of dried plant is dipped overnight in 5 litter water and given for dysentery and worm infection for a week
7.	*Annona squamosa* L.	Sitafal	Annonaceae	Tree	Leaf, seed	Leaf and seed pastes are used externally for tick infection of cattle
8.	*Arisaema tortuosum* (Wall) Schott.	-	Araceae	Herb	Root	Root paste is used externally on wounds and skin diseases
9.	*Azadirachta indica* A. Juss	Neem	Meliaceae	Tree	Leaf	Leaf paste is used externally on wounds and skin diseases
10.	*Brassica juncea* (L.) Czem	Soursa	Brassicaceae	Herb	Fruit	Mustard oil mixed with black ash is directly applied on to the wounds of cattle for wound healing

(Contd.)

S.No.	Botanical name	Local name	Family	Plant type	Parts used	Ethno-veterinary practices
11.	*Bunium persicum* (Baiss.) B. Fedisch.	Kalijira	Apiaceae	Herb	Fruit	Fruit is mixed with garlic, long piper and coriander to make paste and given along with boiled nice for increase lactation of cattle, with preparation is given regularly for one month and also used as an expectorant in cough
12.	*Calotropis gigantean* (L) Dryand	Akanda	Asclepiadaceae	Herb	Leaf	Lear crushed in hot water to prepare paste and raped in the affected parts of the body with help of cloth to treat arthritis and fever in cattle.
13.	*Centella asiatica* (L.) Urban	Thankuni	Apiaceae	Herb	Leaf	Leaves are used in fever and dysentery, also given to the cattle for increasing location
14.	*Cissus quadr angularis*	Harjora	Vitaceae	Climber	Stem	Stem decoction is used externally in wound and dog bite
15.	*Curcuma domestica* Val	Haldi	Zingiberaceae	Herb	Rhizome	50 gm paste is roaly given in constipation, rhizome paste is externally used in skin disease and bone fracture
16.	*Cynodon dactylon* (L) Pers	Durba	Poaceae	Grass	Whole plant	Whole plant paste is externally used in bone fracture, for mdigestion paste is given only
17.	*Cyperus rotundus* L.	Mutha	Cyperaceae	Grass	Root	Crushed root is used in fever
18.	*Echinochola crus galli* (L.) P. Beauv.	Sharmadhan	Poaceae	Grass	Whole plant	Whole plant is given to the cattle for increasing lactation

(Contd.)

S.No.	Botanical name	Local name	Family	Plant type	Parts used	Ethno-veterinary practices
19.	*Eclipta alba* (L.) Hassk	Kesud	Asteraceae	Herb	Leaf	Leaf paste is used as antiseptic and also used externally for wound recovery
20.	*Eleusine indica* (L.) Gaetn.	Ghuradubba	Poaceae	Grass	Stem, inflorescence	Stem with inflorescence is directly rapped on to the swelling leg joint of domestic hen, cock and birds
21.	*Ficus hispida* L. f.	Dumur	Moraceae	Tree	Fruit	5-10 fruits are given daily for recovery from diarrhea
22.	*Hibiscus rosa-sinesis* L.	Joba	Malvaceae	Shrub	Flower bud	Flower buds are crushed and given orally for anoestrus of cattle
23.	*Holarrhena anttdysenterica* Flen	Kruchi	Apocynaceae	Tree	Leaf, Bark	Leaf and bark are used externally for insectbite and wound recovery
24.	*Jatropha curcas* L.	Keron	Euphorbiaceae	Shrub	Leaf	Leaf decoction is applied externally on wound caused by maggots of flies
25.	*Leucus aspera* (Wild.) Link	Swetadron	Lamiaceae	Herb	Whole plant	50 gm of whole plant paste is given orally for ephemeral fever of cattle
26.	*Melia azedarach* L.	Mahaneem	Meliaceae	Tree	Leaf	100 ml leaf decoction is given once a day for a week in loss of appetite
27.	*Mimosa pudica* L.	Lajjyabati	Mimosaceae	Herb	Root	10 gm of root paste is given in recurrent fever of cattle
28.	*Musaparadisiaca* L.	Kala	Musaceae	Shrub	Fruit	10-15 fruit are given in constipation problem, also given with fodder to increase lactation

(*Contd.*)

S.No.	Botanical name	Local name	Family	Plant type	Parts used	Ethno-veterinary practices
29.	*Nicotiana rustica* L.	Sadapata	Solanaceae	Tree	Leaf	Dry leaf powder mixed with naphthalene is used to remove maggots from wounds of domestic cattle
30.	*Ocimum sanctum* L.	Tulsi	Lamiaceae	Herb	Leaf	10-15 gm leaf paste is given daily for a week for healing from cough and cold
31.	*Piper betel* L.	Pan	Piperaceae	Climber	Leaf	2-3 leaves are given orally in diarrhea of cattle
32.	*Polygonum hydropiper*	Biskatali	Polygonaceae	Herb	Whole plant	Whole plant is crushed with water and is used on the skin of cattle to remove ticks once in a month
33.	*Ricinus communis* L.	Rehri	Euphorbiaceae	Shrub	Leaf	Leaves given to the cows for increasing lactation
34.	*Saccharum officinarum* L.	Aakh	Poaceae	Shrub	Stem	Molasses prepared from sugarcane is applied on tongue infection
35.	*Sasbania grandiflora* (L.) Poir	Bakful	Leguminosae	Tree	Leaf, flower	Leaves and flowers are used as supplementary food diet of cattle with fodder
36.	*Solanum xanthocarpum* Schard. & Wendl.	Kantikari	Solanaceae	Herb	Whole plant	100 ml of whole plant decoction is given orally twice a day for cough and cold
37.	*Spermacoce hispida* L.	Refugilata	Rubiaceae	Climber	Whole plant	Whole plant is used to treat dysentery in cattle
38.	*Streblus asper* Lour	Sheora	Moraceae	Tree	Leaf	Leaves are used to treat dysentery in cattle
39.	*Tagetes erecta* L.	Genda	Asteraceae	Herb	Whole plant	Whole plant is crushed and the paste is applied externally on foot ulcer

(Contd.)

S.No.	Botanical name	Local name	Family	Plant type	Parts used	Ethno-veterinary practices
40.	*Terminallia bellirica* (Gaertn) Roxb.	Bahera	Combretaceae	Tree	Leaf	Leaves are given to the cows for increasing lactation
41.	*Trichosanthes biracteata* (Lamk) Voight (Syn *Trichosanthes palmala*)	Makal	Cucurbitaceae	Climber	Root	25 gm root paste is given daily for a fortnight for bronchial problem of cattle
42.	*Vernania cinera* (L.) Less	Siyalmutra	Asteraceae	Herb	Leaf	Leaf paste is used externally in cuts and wounds
43.	*Vignamunga* (L.) Hepper	Mugdal	Fabaceae	Herb	Seed	Seed are given to the cows for increasing lactation
44.	*Vitex negunab* L.	Nisinda	Verbenaceae	Tree	Leaf	Leaves with rhizome of *Curcuma longa* are crushed together in ratio of 2:1 and applied externally to reduce pain; dry leaves are used as bed when domestic hen, cock etc. are attacked by external parasites
45.	*Zingiber officinale* Rose	Ada	Zingiberaceae	Herb	Rhizome	10-15 gm of paste is given for constipation and indigestion problem

SUMMARY

- Tripura a small state of North-eastern India is rich in assemblage of medicinal plants which are used by the local people for the treatment of different ailments of domestic animals.
- Based on the result of a survey in the different parts of Tripura, 45 species of ethno-veterinary medicinal plants were documented.
- The recorded plant species include herbs (42.22%), trees (26.67%), shrubs (13.33%), climbers (8.89%), and grasses (8.89%).
- Leaves are used as principal raw material in 41.7% medicine preparations followed by whole plant (18.8%) and roots (8.3%) and so on.
- Unscientific collection of raw materials, developmental activities and population pressure in the forested areas are the main threats to the medicinal plant diversity of the state which need urgent implementation of sustainable conservation strategies.

CONCLUSION

The ethno-botanical wisdom has not been sufficiently explored till now and there are a lot of unknown plant species which are yet to be explored for their medicinal values. Preservation of indigenous knowledge of plants used in traditional health care is very important. Expanding population growth, rapid industrialization and pollution are important reasons for the destruction of biodiversity. Therefore, the need of proper action plan is emphasized for the sustainable conservation of year old practices of ethno-veterinary plants and practices.

ACKNOWLEDGEMENT

Authors are grateful to the local healers and ethno-veterinary medicine men for sharing their traditional knowledge. Authors are also thankful to Dipankar Kishore Sinha for his assistance in field studies.

References & Recommended Reading

Bandyopadhyay, S. and Mukherjee, S. K., 2005. Ethnoveterinary medicine from Koch Bihar district, West Bengal. *Indian Journal of Traditional Knowledge* **4**(4): 456-461.

Bharati, A. K. and Sharma, B. L., 2009. Studies on Ethnoveterinary uses of Plant Resources of Sikkim, *The Indian Forester* **135** (5): 691-696.

Bisht, N., Pande, P. C. and Tiwari, L., 2004. Ethnoveterinary practices of Kapkot block of Bageshwar District, Uttaranchal, *Asian Agri History* **8**(4): 209-314.

Borthakur, S. K. and Sharma, V. K., 1994. Ethno veterinary medicine prevalent among Nepalis of Assam. Paper presented at the *Fourth International Congress of Ethnobiology*, held at the National Botanical Research Institute, Lucknow, India, 17- 21 November, 1994.

Gaur, R.D., Bhatt, K.C. and Tiwari, J.K., 1992. An Ethnobotanical study of Uttar Pradesh Himalaya in relation to Veterinary medicines. *Journal of Indian Botanical Society* **72**:139-144.

Harsha, V. H., Shripathi, V. and Hegde, G. R., 2005. Ethnoveterinary practices in Uttara Kannada district of Karnataka. *Indian Journal of Traditional Knowledge* **4**(3): 253-258.

Iswar, R. K., 1981. Traditionally important medicinal plants and folklore of Uttaranchal Himalaya for animal treatments. *Journal of the Scientific Research on Plants and Medicine* **2**: 61-66.

Jain, S. K., 1966. *Ethno botany In Human welfare*. Deep Publications, New Delhi.

Kiruba, S., Jeeva, S. and Dhas, S. S. M., 2006. Enumeration of ethnoveterinary plants of Cape Comorin, Tamil Nadu. *Indian Journal of Traditional Knowledge* **5**(4): 576-578.

Kshirsagar, R. and Upadhyay, S., 2009. Free radical scavenging activity screening of medicinal plants from Tripura, Northeast India. *Natural Product Radiance* **8**:117-122.

Kumar, A., Pandey, V. C. and Tewari, D. D., 2012. Documentation and determination of consensus about phytotherapeutic veterinary practices among the Tharu tribal community of Uttar Pradesh, India. *Tropical Animal Health and Production* **44**: 863-872.

Maheswari, J. K., Kalakoti, B. S. and Brij, L., 1986. Ethnomedicine of Bhil tribe of Jhabua district, M. P. *Ancient Science of Life* **5**(4): 255-261.

Mathias, E., 2004. Ethnoveterinary medicine: harnessing its potential. *Veterinary Bulletin* **74**(8): 27–37.

McCorkle, C., 1986. An introduction to ethnoveterinary research and development. *Journal of Epidemiology* **6**: 129-149.

Mishra, D. 2011. Ethnoveterinary practices and use of herbal medicines for treatment of skin diseases in cattle: A study in Polsara Block, Ganjam District, Orissa, India. *Veterinary World* **4**(6): 250-253.

Mishra, S., Sharma, S., Vasudevan, P., Bhatt, R. K. Pandey, S. Singh, M., Meena, B. S. and Pandey, S. N., 2010. Livestock feeding and traditional healthcare practices in Bundelkhand region of Central India. *Indian Journal of Traditional Know*ledge **9**(2): 333-337.

Nigam, G. and Sharma, N. K., 2010. Ethnoveterinary plants of Jhansi district, Uttar Pradesh. *Indian Journal of Traditional Knowledge* **9**(4): 664-667.

Pandey, P. C. and Joshi, G. C., 1999. Kumaun Himalaya ke Pashu Chikitsa aadi main paltu pashuon ka upyog (in Hindi). *Ethnobotany of Kumaon Himalaya* (Pande P.C. et al., Eds). Scientific Publishers, Jodhpur, India, pp113-123.

Pandit, P. K., 2010. Inventory of ethno veterinary medicinal plants of Jhargram division, West Bengal, India. *The Indian Forester* **136** (9): 1183-1194.

Priyadarsan, S. S., 1991. Herbal Veterinary medicines in an Ancient Sanskrit work 'Garuda Purana'. *Ethnobotany* **3**: 83-87.

Reddy, K. J. and Sudarsanam, G., 1987. Plants used as veterinary medicine in Chittor district of Andhra Pradesh, India. *International Journal of Crude drug Research* **25**: 145-152.

Salave, A. P., Reddy, P. G. and Diwakar, P. G., 2011. Some reports on ethnoveterinary practices in Ashti areas of Beed District (M. S.) India. *International Journal of Applied Biology and Pharmaceutical Technology* **2**(2): 69-73.

Salave, A. P., Sonawane, B. N. and Reddy, P. G. D., 2012. Traditional ethno-veterinary practices in Karanji Ghat areas of Pathardi Tahasil in Ahmednagar district (M. S.) India. *International Journal of Plant, Animal and Environmental Sciences* **2**(1): 64-69.

Samal, P. K., Shah, A., Tiwari, S. C. and Agarwal, D. K., 2002. Indigenous animal health care practices and their relevance to bio resources conservation in Indian Central Himalaya. *International Journal of Sustainable Development and World Ecology* **8**: 167-178.

Samal, P. K., Shah, A., Tiwari, S., Mehra, B. and Agarwal, D. K., 2003. Indigenous animal health care practices in India Central Himalaya: empirical evidences. *Indian Journal of Traditional Knowledge* **2**(1): 40-50.

Satapathy, K. B., 2010. Ethnoveterinary practices in Jajpur district of Orissa. *Indian Journal of Traditional Knowledge* **9**(2): 338-343.

Sebastine, M. K., 1984. Plants used as veterinary medicines, galactogogues and fodder in four areas of Rajasthan. *Journal of Economic and Taxonomic Botany* **5**: 785-788.

Sebastine, M. K. and Bhandari, M. M., 1984. Some plants used as veterinary medicine by Bhils. *International Journal of Tropical Agriculture* **2**: 307-310.

Thomas, B. A., Rajendran, V., Aravindhan, and Sivalingam, R., 2011. Ethnoveterinary medicines of tribe Paniyars in Kerala, India. *International Journal of Biological Technology* **2**(2):72-75.

Tiwari, D., Pande, P. C. and Tiwari, L., 2007. Ethnoveterinary Herbal Medicine of Dwarahat area of Central Himalayas. *The Indian Forester* **133**(3): 379-390.

Tiwari, L. and Pande, P. C., 2004. Traditional Veterinary Practices in South-eastern part of Chamoli District, Uttaranchal. *Indian Journal of Traditional Knowledge* **3**(4): 397-406.

Tiwari, L. and Pande, P. C., 2010. Ethnoveterinary medicines in Indian perspective: Reference to Uttarakhand, Himalaya. *Indian Journal of Traditional Knowledge* **9**(3): 611-617.

Vidyanath, J., Chowdhury, U. N. and Saraswati, K. C., 1991. Botanical aspects of an ethno veterinary prescription in Mithila, North Bihar (India). *Ethnobotany* **3**:101-104.

Chapter – 8

Indigenous Traditional Knowledge Associated with Fermented Fish Products of North East India

A.U. Muzaddadi

INTRODUCTION

The fermented products probably came into existence at least before the British Era in Northeastern states of India [before 1824, The Yandabu Agreement between British Govt. and Burma (now Myanmar)] when the British first introduced salt in the region. Interestingly, in all the fermented fish products of this region no salt is used as ingredient which is a major difference with the other fish based fermented fish products of Southeast Asian countries such as *Nuac-mum* of Vietnam, *Budu* of Malaysia, *Patis* of the Philippines etc. Even after the British Era, people used to treat salt as a highly valued and scarce commodity and they used to take an alternative commodity, for example *khar* in Assam, made from banana plant or papaya. Hence, they could not afford spending salt in fish preservation and they were bound to preserve fish in their own cheaper and easier way. As Saisithi (1994) suggested, it is quite natural that the simple drying might have prolonged due to frequent rainfall resulting in high humidity in the atmosphere particularly during the peak fishing seasons, i.e., monsoon and post-monsoon from May to September. Consequently, the experienced fishers had to find out a method through which they could preserve the instant heavy catches comprising of very less valued small fishes like *Puntius* spp., *Amblypharyngdon* sp., *Trichogaster* spp., *Channa* spp., *Mystus* spp., *Botia* spp., *Danio* spp., *Rasbora* spp., *Chanda* spp., *Nandus* sp., etc. for the consumption and sale in the dry seasons (November to April) when raw fresh fish used to be scares in the market. Due to the unavailability of ice and good road communication facilities, fishers could not even transport this highly perishable catch to the distant markets where they could have received a profitable price. As a matter of fact, the fishers would never earn good price in the village markets through instant sale of these fish.

Furthermore, the people, perhaps, wanted to change the taste and flavor of their daily rice-fish dish, which had become bland and monotonous. As a result, they invented the methods of preparation of such fermented fish products. For instance, people must have stored the semi-dried fish in earthen vats (locally called *mutkas*) for months together which eventually allowed the fermentation process to set in and resulted in a semi-pasty and product with strong smell "*seedal*" in Assam. People gradually acquired the taste and aroma of such fermented products and especially the disadvantaged economic conditions of the people, perhaps, created compulsion for accepting these products. The primary objectionable smell of these products was taken care of while preparation of the culinary dishes by adding hot spices including green chili, ginger, garlic, onion, red pepper, salt and oil and by frying or boiling with subsequent grinding into pastes.

The use of small economic varieties of fish in fermented fish products and their popularity in Northeast may be due to several reasons such as [1] Highly perishable nature of such fish did not allow transport to the distant markets for better price, thus compelling fishers to go for preservation, [2] Variability of size and shape of individual species in the whole catch, however, species wise uniformity in size such as *Puntius* spp. made it easier to sort and segregate this particular species for processing for preparation of seedal. [3] The fish oil required for product processing could be extracted sufficiently from the same fish, hence no dependency on other oil source and a low cost product. [4] The other options like preparation of silage were not known to the fishers. [5] Simple sun drying was difficult enough due to frequent rain, infestation by pests and creation of unpleasant atmosphere in the courtyard. [6] Believing such products to be of medicinal value. [7] The culinary preparations of these products used to be efficient appetizer and taste enhancer to serve as condiment with rice.

METHODOLOGY

The information was collected through extensive field study and survey. The study area was Nagaon district of Assam, Imphal West district of Manipur, Ri-Bhoi and East Khasi Hills districts of Meghalaya and West Tripura district of Tripura. Three production centers from each district were surveyed and the persons actively engaged in the production were interviewed. From the many variations of the preparation procedures adopted in the study region, a common method of preparation of the product has been worked out after eliminating insignificant deviations having little bearing on the finished product quality.

RESULTS AND DISCUSSION

Seedal

Seedal is a pasty and solid product in which the shape of the fish (*Puntius* spp.) remains almost intact (Fig. 1). Instead of salt, as in many other indigenous fermented fish product, extra oil is added here during product preparation. Hence, rancidity is a must in *seedal.*

Puntius spp. is exclusively used. The rural fishers procure them from beels and rivers. Commercially it is prepared from dried *Puntius* spp. which can be grouped into four groups in Assam. The first group is *Desi puthi*, produced locally and considered to be the best quality raw material for s*eedal* production. The dried *Puntius* spp. belonging to the other groups come from Madras, Luckhnow and Gurakhpur, hence named as *Madrasi Puthi*, *Luckhnowi Puthi* and *Gurakhpuri Puthi.* All these varieties are procured from Jagiroad Fish Market (about 50 km east to Guwahati City, Assam). As they are imported from distant places through truck loadings, their quality deteriorates during transportation due to the unpredictable delay. However, due to the short supply of Desi puthi, the imported varieties are used by the commercial producers by compulsion which made the producers able to produce *seedal* throughout the year and maintain sustained supply to meet the high market demand. Moreover, it made them independent of the highly erratic availability of Deshi puthi. Some commonly used species of *Puntius* spp. are given in Table 1.

Table 1: The major species of Puntius used in the preparation of seedal/shidal

Sl. No.	Scientific Name	English common name	Assamese local name	Normal size range (cm)
1.	*Puntius chola* (Hamilton, 1822)	Swamp barb	Doloni puthi	2-5
2.	*P. sarana* (Valenciennes, 1842)	Olive barb	Seni puthi	3-8
3.	*P. sophore* (McClelland, 1839)	Pool barb	Soru puthi	3-6
4.	*P. ticto (*Hamilton, 1822)	Two spot barb	Tita puthi	3-8

*Mutka*s are round bottom and narrow necked specially made earthen pots, the capacity of which ranges from 8 to 40 kg. Mutkas can be used for preparation of many batches of product until they break. Older the *mutka*s better the product quality and lesser the cost of production, the reason being the older *mutka*s absorb very less amount of oil and provide very good gas barrier property.

Fish oil extracted from the entrails of *Puntius* spp. or other fresh water fish through an indigenous crude method is used by the rural producers. Oil is used to smear the *mutka*s, perhaps to increase gas barrier property by filling the minute pores in the earthen pot. Although fish oil is treated as the best oil for smearing mutka, for the commercial production vegetable oil such as mustard oil is used as an alternative of fish oil due to its easy availability and inexpensiveness in comparison to fish oil.

Semi-dried or fully dried fish are often crushed into powder and a paste is made from it which is used to fill in the mouth portion of *mutka*s.

Any broad leaf of plants such as *Colocassia* or banana leaves, bottle-guard leaves or a piece of newspaper is used to seal the cover paste temporarily.

A thick mud is prepared from humus rich and fine coarse soil, which is used to seal the mouth.

THE PROCESS

This procedure is popular among the rural fishers where raw fresh fish are available in plenty. Fishers extract fish oil from entrails of *Puntius* spp. and make use of it in *seedal* preparation (Flow Chart 1). This naturally evolved procedure is most economic and climate friendly and mostly practiced in the backyard of fishers. Although the s*eedal* produced by this method is considered to be the best, the commercial producers do not normally follow this method, perhaps due to inconsistent supply of raw material.

Fig.1: *Seedal* of Assam

Indigenous Procedure of *Seedal* Preparation

Raw material (Raw fresh fish *Puntius* spp.)

↓

Descale, gut and wash

↓

Semi-dry in the sun (about 50% of the total moisture is retained in fish)

↓

Pack the semidried fish in oil-processed *mutkas*[1]

↓

Fill the mouth portion (from the neck to rim of the *mutka*) with the cover-paste

↓

Cover the paste with the cover-leaf and take out the *mutka* from the ground.

↓

Keep it in a sheltered place for 3-4 days as such

↓

Put a layer of thick mud on the paste after removing the cover leaf

↓

Keep in a dry and sheltered place. If cracks appears on the mud-seal, put another layer of mud layer on the previous one and repeat it till a crackless seal is attained

↓

Now keep the *mutka* undisturbed for 3-4 months

↓

Remove the mud and putrefied paste carefully and take out the final product one layer after another

Shidal

Shidal, a seedal like product, is prepared in Tripura (one of the states of Northeast India) mainly from two species of fish such as *Puntius* spp. (Fig. 2) and *Setipinna phasa*. Here completely dry fish are used and vegetable oil is used instead of the fish oil for smearing of mutka. The dried fish is further dried under sun after a thorough sorting from the whole lot. The uniform sized fleshy fish (not infested by maggots) are dried to reduce the moisture level to about 15 percent. The fish are then washed thoroughly and dipped into water 10-15 minutes in order to make them slightly soft. These fish are then dried under shade overnight spreading under a drying shed. In the very next morning, the mutkas are packed by putting the fish in layers in a pre-processed mutka very tightly. Mutkas are processed by smearing with mustard oil and repeatedly dried under sunlight until they become saturated with oil. These mutkas are kept under ventilated shade for 3-4 months for fermentation. The fermentation time varies depending on the texture of dry fish which is used as raw material.

The advantages of this method may be listed as

- No dependency on seasonally available fresh fish, thus shidal can be prepared round the year
- The large scale commercial production is possible due to a sustained supply of raw material.
- The fish oil, with short supply, is replaced with vegetable oil resulting in large scale production

Although all types of small carps belonging to the genera *Puntius* are used for preparation of quality products, best quality product is prepared from the matured fish of approximately 3-6 cm in length of the species given in Table 1. The rural fishers believe that the fish before bearing eggs are best suited for seedal or shidal preparation.

Shidal is prepared in commercial scale to meet the demand in the Northeast states of India. As a matter of fact, the commercial producers modified the traditional method to suit large scale production. Hence, they compromised with the traditional originality of the product and incorporated those raw materials which have a constant supply for year round production along with reduction of normal fermentation time. Consequently, the product (shidal) retained nearly the original shape of the raw material containing less organoleptic properties compared to seedal.

Fig.2: *Shidal* of Tripura

Numsing

Numsing is a semi dried and semi smoked paste like product which is prepared by mixing fish, petioles of arum and spices (Fig. 3). Small economic fish species of assorted size abundantly available during the south-west monsoon season (June-September) in the study region (Table 2) are used for preparation of the product together with petioles of Elephant ear (*Alocasia macrorrhiza*) belonging to the family Araceae, locally known as *Mana Kasu*. It is a herbaceous perennial plant having broad leaves and a large rizhome on or just below the ground. The fish are dressed to remove gills, scales and intestine and washed with clean water. The edible part of arum is sorted out, peeled and sliced using kitchen knife (*kotaari*). Fish are dried on a specially made bamboo rack. It is made using bamboo with a height of 3-4 feet and having a fireplace at its bottom. Fish are spread on it and heated with nearly a smokeless fire that is created by burning fire woods or dry bamboo at the fireplace mostly during night hours. Drying and smoking is continued with mild flame until the fish become moderately hard. Arum slices are sundried for 1 day. The flame dried fish and sundried arum slices are mixed together in the ratio of 4:1. Other spices such as red pepper, green chili, ginger and garlic may also be used according to the taste. However, no salt is used. A foot operated elongated traditional wooden grinder or huller (locally called *Dhekee*) is used to grind the mixture. *Dhekee* is a household apparatus in Assam that is usually used for hulling rice and grinding spices. In households that lack *dhekee*, a smaller hand operated traditional wooden grinder/huller pot with a wooden thrasher (*uraal*) (involves higher drudgery) is used for grinding and mixing the ingredients. Grinding continues until the mixture becomes coarse paste. The ground mixture, which is by now a coarse paste, is stuffed tightly in a bamboo container (*bahor chunga*) keeping headspace of about 10-15 cm for subsequent sealing. The container is a single internode of immature bamboo (*Bambusa* spp., less than one year old) that is made by retaining one node and excluding the other. These tubes also serve as the primary packaging material for the product. The stuffed bamboo containers are now re-stuffed with the raw leaves of bladder fern (or fragile fern) locally known as *Biholongoni* that belongs to the genus *Cystopteris* sp. The leaves of this plant are washed and stuffed keeping headspace of about 5-10 cm. Dry rice straw is used as secondary sealing material. Fine straws are cleaned, dried properly and then pushed hard inside the bamboo container in such a way that a headspace of about 2-5 cm is retained. Finally the bamboo tube is sealed air-tight with clay so that no air or insects can penetrate it. Fine clay soil is crushed and mixed with water to make thick clay. The sealed bamboo containers are then hung about 2-3 feet above the traditional mud-oven in the kitchen or a fireplace whereupon the *Mising* people cook food as outdoor kitchen. After allowing a fermentation period of about 30 days in this manner, which allows an intermittent heating, the product becomes ready for consumption. The product is usually allowed to be inside the fermenting bamboo container itself for further storage. Small quantities are taken out from the container for cooking after removing all the sealing materials. For convenience, the final product is sometimes taken out from the container by cutting it open and kept in a glass jar for further use. It has a shelf life of about 2-3 years.

Fig. 3: *Numsing* of Assam

Numsing is consumed after steam cooking or after preparing some curry along with vegetables. For steam cooking, required quantity of *numsing* is taken out from the bamboo tube and mixed with chili and few other spices such as onion, garlic, turmeric and ginger. All ingredients are mixed properly and wrapped with fresh raw leaves of a certain wild shrub that has broad and slender leaves, belongs to ginger family Zingiberaceae and locally known as *tora gos*. The wrapped packet is tied with cotton threads and cooked in steam along with rice. This mixture is then ground to a paste or chutney and eaten with rice.

A *numsing* curry with vegetables such as ridge gourd and leafy vegetables such as spinach, mustard leaves, pumpkin leaves, arum leaves, etc is also prepared. The product is mixed with ingredients of curry and cooked in a curry-pan with sufficient amount of gravy. It gives a sweet flavour and a characteristic odour to the curry.

Table 2: Small value fish used in *numsing* preparation

English name	Scientific name	Common /Local name	Normal size range (cm)
Indian carplet	*Amblypharyngodon microlepis* (Bleeker, 1853)	*Muwa*	2-5
Mola carplet	*A. mola* (Hamilton, 1822)	*Muwa*	2-5
Aspidoparia	*Aspidoparia morar* (Hamilton, 1822)	*Baralia*	4-12
Spotted mural	*Channa punctatus* (Hamilton, 1882)	*Goroi*	5-20
Indian hatchet chela	*Chela cachius* (Hamilton, 1822)	*Chelekona*	3-6
Zebra danio	*Danio rerio* (Hamilton, 1822)	*Dorikona*	1-5
Giant danio	*Devario aequipinnatus* (McClelland, 1839)	*Seldorikona*	3-10
Flying barb	*Esomus danricus* (Hamilton, 1822)	*Dorikona*	3-10
Indian river shad	*Gudusia chapra* (Hamilton, 1822)	*Koroti*	5-15
Guntea loach	*Lepidocephalus guntea* (Hamilton, 1822)	*Gutum/botia*	2-15
Indian glass barb	*Laubuca laubuca* (Hamilton, 1822)	*Laupati*	5-15
Spotted spiny eel	*Macrognathus puncalus* (Cuvier, 1832)	*Tora/turi*	8-30
Peacock eel	*M. aral* (Cuvier, 1832)	*Tora/turi*	8-30
Bengala barb	*Megarasbora elanga* (Hamilton, 1822)	*Eleng*	5-15
Banded dwarf catfish	*Mystus vittatus* (Hamilton, 1822)	*Singora/tingora*	3-10
Swamp barb	*Puntius chola* (Hamilton, 1822)	*Doloni puthi*	2-5
Olive barb	*P. sarana* (Valenciennes, 1842)	*Cheni puthi*	3-8
Pool barb	*P. sophore* (McClelland, 1839)	*Soru puthi*	3-6
Two spot barb	*P. ticto* (Hamilton, 1822)	*Tita puthi*	3-8
Slender rasbora	*Rasbora daniconius* (Hamilton, 1822)	*Dorikona*	3-10
Gangetic scissortail rasbora	*R. rasbora* (Hamilton, 1822)	*Dorikona*	5-10
Finescale razorbelly minnow	*Salmophasia phulo* (Hamilton, 1822)	*Selkona*	4-12
Honey gourami	*Trichogaster chuna* (Hamilton, 1822)	*Becha*	3-7
Banded gourami	*T. fasciata* (Bloch & Schneider, 1801)	*Kholihona*	5-12
Thick lipped gourami	*T. labiosa* (Day, 1877)	*Ranga kholihona*	4-9
Dwarf gourami	*T. lalius* (Hamilton, 1822)	*Kholihona*	5-8

Different ethonobotanical resources used in traditional foods are based on the location specific demand, culture, economy, ethnicity, food habit and overall needs (Jeyaram et al., 2009). *Numsing* is normally prepared during monsoon season when the trash fish and arum are available in the villages in plenty. Trash fishes are highly perishable in nature due to neutral pH content of fish and easy microbial attack. Moreover, in the monsoon season high valued big fish become comparatively cheaper due to increased supply in the market. Consequently people opt for bigger fish for fresh consumption and there is a glut of small fish. These fish naturally find way to be preserved for the seasons such as winter and spring (December to April) when fish has a short supply. Hence, the *Mising* tribes, who are habituated to dwell in nearby rivers and *beels* (typical freshwater lakes/marshy areas found in the Brahmaputra valley), have developed this method of preservation and value addition to low cost fish by the urge of natural instinct.

The drying method of raw fish on the bamboo platform is another marvelous invention which is excellently suitable for monsoon season. During consistent cloudy days of monsoon, the method of traditional sun drying of fish becomes completely impracticable. In addition, this method of drying has several disadvantages such as (1) unpredictability of sunshine; (2) operation hours during peak day time only which is valuable for other farm activities; (3) high probability of insect infestation during drying; (4) attack by predators like birds, cats and rats; (5) bacterial and fungal attack on the fish and finally (6) slow rate of drying. The new method of fish drying amicably solves all these problems and it becomes worth replacing the traditional sun drying of fish. Besides, this accelerated method of drying renders cooked flavour to the product.

The second major ingredient of the product is *mana kosu* which is rich in starch and thus supplements carbohydrate in *numsing*. Fish being very poor in carbohydrate content (0.5% in w/w basis), this ingredient sufficiently compensates it in the product. It also contributes to increase the viscosity of the product and thus increasing the thickness. People believe that providing an exceptional flavour, *mana kasu* has some preservative action too.

The bamboo tube used is an organic packaging material. The raw bamboo protects the product from getting burnt while heating and also gives additional flavour to the product. Having a very hard surface it protect's the product from insects, rats and cats. The tube can be easily opened by splitting it apart or by cutting the node at the bottom.

The fern leaves have some insect repellant properties and it protects the product from insect infestation. It is also believed that the unpleasant smell of fish which would attract insects is absorbed by these leaves to some extent.

Other sealing materials such as rice straw probably would work as insulator and spongy cushion to accommodate any increment in volume of the product or gas production. On other hand, clay becomes hard on drying and gives a good protection against insect infestation and attack by rats. It also forms nearly an airtight seal protecting leakage of foul smelling gases from fish inside the bamboo tube.

The product is at present used for own consumption. Since most of water bodies, wetlands and low lying paddy fields in the study region dries up during the pre-monsoon season (March- May), availability of small economic fish in the study region is drastically reduced. Since most of the people in the study area are socio-economically underdeveloped, they cannot afford to buy fish from the open market during this period. Thus, the traditional system of processing fish which are abundantly available during rainy season and preserving these fish for the consumption during the scares seasons, is a useful means of avoiding malnutrition based on the traditional wisdom.

Of late, there has been a resurgent of interests in the ethnic foods prepared by different tribes/communities of the region as seemed from the food festivals organized by different Govt. and Non-govt. organizations from time to time, mainly because most of the ethnic foods prepared by the tribals usually contain less/ negligible edible oil including potentially harmful Low Density Lipoprotein (LDL), harmful preservatives and colouring agents. Moreover, being a salt free product, it is a health food for heart patients with high blood pressure.

The product is a resultant of combination of three processes such as drying, smoking and fermentation which makes the product very unique among all processed fish products. Moreover, the addition of arum petioles undoubtedly would increase the dietary fibre content of the product which is a major concern in modern health food formulation. In fact, the use of immature bamboo containers for stuffing and storing the product, provides a distinct raw bamboo aroma to it making it more special for the tribes/communities of the region. In addition, the product has a reasonably long storage life of at least 2 years.

The processing method is undoubtedly an environment friendly method which utilizes the available resources for preservation and value addition to low cost fish. The ingredients, packaging and sealing materials surely would have some unknown biochemical roles to make the product special. The authors are undertaking a small project for conducting basic research on this product.

Ngari

Sun dried non-salted dry fish called Phoubu (*Puntius sophore*) is brought from Assam and Bangladesh and fermented by traditional ways in valley region of Imphal and surrounding areas. Dry *Puntius* spp. is collected during October to January @ Rs 100-300/kg. Before packing in oil-processed earthen pots, fish is washed with water using porous bamboo baskets and allowed to drain the water overnight. Next day morning, water-soaked fish are pressed by legs keeping under gunny bags and the excess water is removed. In this process the head & bones of fish are broken. Oil released from head during pressing is believed to cause initiating fermentation. Earthen pot (capacity of 45-50 kg) inner surface is smeared with mustard oil. For new pots, 8-10 times oil coating are required with interval of 7-10 days. In old pots, only one coat is sufficient for quality fermentation. Oil smearing, perhaps, blocks the minute pores in the earthen pots and an anaerobic environment inside the chamber is created. The washed and pressed dry fish is packed tightly inside the pot by pressing with foot. Only the experts are engaged for this purpose. After packing, the mouth portion of the pots are sealed with fish scales, oil slurry, cow dung slurry, polythene sheet and finally with mud. These packed pots are kept in dark up to 6-12 months at room temperature. After about 3 months of fermentation the pots are opened by removing the sealing material and the ready *ngari* is taken out.

It is consumed after preparation of dishes with bamboo shoots and other fermented vegetable products. It is consumed as condiments with rice by Manipuri people in particular and even popular in Myanmar.

Hentak

Hentak is a traditional fermented fish paste, prepared in Manipur of Northeast India. During indigenous preparation process, the sun dried *Esomus danricus* (a small weed fish) are crushed into power. The petioles of *Alocasia macrorhiza* are cut into pieces washed with water and exposed to sunlight for an hour. An equal weight of the cut pieces is then crushed along with fish powder to make a paste. Small balls are prepared and put in earthen pots and stored. After two weeks of fermentation, it is ready to use and are preserved. These become hardened on keeping for few months and are then propounded to paste with a little water and stored as balls for reserve food.

Tungtap

Tungtap is a fermented fish prepared and consumed by *Khasi* and *Jayantia* tribes of Meghalaya of Northeast India. The product is prepared from some small varieties of fish including *Puntius* spp. and *Danio* spp. Dry fish are washed properly and dried under sun for 3-4 days so that the surfaces are just dried. Fish are mixed with fish oil, normally extracted from the viscera of fish which gives additional aroma, flavor and taste. Fish are then tightly packed in oil-treated earthen pot which ensures a semi-anaerobic environment. The mouth portion is sealed with crushed fish paste and mud. Earthen pots are stored in room temperature (18-28°C) for 3-6 months until use. It is consumed after preparing curry along with rice.

CONCLUSION

Traditional fermented fish products of Northeast India deserve to attract more attention from scientific communities. Present article is expected to provide some basic information for detailed scientific studies on the products in near future. Since the products are prepared in a ecofreindly environment with all natural ingredients, there is every possibility presence of neutracuticals and healthful metabolites in these products. The belief that many of these products possibly have medicinal value such as anti-malarial property of shidal and ngari, is yet to be proven scientifically. We are, at present, actively engaged in such research and desirous to give an identity to these products in the global scenario.

ACKNOWLEDGEMENTS

The author is highly thankful to the Muslim Fisher Community of Nagaon district of Assam and Hindu fisher community of Tripura for providing the valuable information about the products. The author is also thankful to Central Agricultural University for providing basic assistance in conducting the research.

References & Recommended Readings

Choudhry, D., Sahu, J. K. and Sharma, G. D., 2011. Bamboo shoot based fermented food products: a review, *J. Sci. Ind. Res.* **70**: 199-203.

Dutta, B. K. and Dutta, P. K., 2005. Potential of ethnomedicinal studies in Northeast India: An overview, *Indian Journal of Traditional Knowledge* **4**(1): 7-14.

Jeyaram, K., Singh, T. A., Romi, W., Devi, R. A., Singh, W. M., Dayanidhi, H., Singh N. R. and Tamang J. P., 2009. Traditional fermented foods of Manipur, *Indian Journal of Traditional Knowledge* **8**(1): 115-121.

Mao, A. A. and Odyuo , N., 2007. Traditional fermented foods of Naga tribes of Northeastern India, *Indian Journal of Traditional Knowledge* **6**(1): 37-41.

Murugkar, D. A. and Subbulakshmi, G., 2006. Preparation techniques and nutritive value of fermented foods from the khasi tribes of Meghalaya, *Ecol. Food and Nutrition* **45**: 27-38.

Muzaddadi, A. U. and Basu, S., 2012. An accelerated process for fermented fish (seedal) production in Northeast region of India, *Indian J. Animal Sc.* **82**(1): 98-106.

Muzaddadi, A. U. and Basu, S., 2003. Microbiological and sensory changes during preparation of seedal- A fermented fish product, in *Seafood Safety*, edited by P K Surendran, P T Mathew, N Thampuran, V N Nambiar, J Joseph, et al. (Society of Fisheries Technologists (India), Cochin) 35-40.

Muzaddadi, A. U. and Basu, S., 2003. *Seedal*-an indigenous fermented fishery product of Northeast India, *Fishing Chimes* **23**: 30-32.

Muzaddadi, A. U. and Basu, S., 2012. *SHIDAL* -A Traditional fermented Fishery product of North east India, *Indian Journal of Traditional Knowledge* **11**(2): 323-328.

Pope, C. G and Stevens, M. F., 1939. The determination of amino nitrogen using a copper method, *J. Biochem.* **33:** 1070-1076

Saisithi, P., 1994. Traditional fermented fish: fish sauce production, in *Fisheries Processing-Biotechnological Applications* edited by Martin, M. A., Chapman and Hall, 111-131.

Singh A., Singh, R. K. and Sureja, A. K., 2007. Cultural significance and diversities of ethnic foods of Northeast India, *Indian Journal of Traditional Knowledge* **6** (1): 79-94.

Thapa, N., Pal, J. and Tamang, J., 2004. Microbial diversity in ngari, hentak and *tungtap*, fermented fish products of North-East India, *World J. Microbiol. Biotech.* **20:** 599-607.

Chapter – 9

Traditional Ethnic Jewellery of Kashmiri Fisherwomen

Rizwana, T. H. Bhat, M. H. Balkhi and Poonam Sharma

INTRODUCTION

Jewellery has been a favorite form of adornment since time immemorial in all parts of the worlds and Indian is not an exception. The significance of jewellery in the country is evident from the fact that on many auspicious occasions, jewellery forms a part of gifts. Jewellery has not only been considered for the purpose of adoration, but it can also be sold to overcome the economic crisis. This way, jewellery also serves the purpose of insurance, which can be depended upon. To accentuate their feminine beauty, women use jewellery made of gold, silver and diamonds. Traditionally, jewellery has always been linked with wealth, power and status.

Kashmir or Kashir occupies a central and strategic position in Asia being surrounded by Russia, China, India, Tibet, Pakistan and Afganistan. It is situated in the lap of Himalayas (Kak, 2002). The Himalayan valley i.e the valley of Kashmir came into existence out of a vast lake surrounded by the Himalayan mountains. Kashmir is renowned throughout the world for its beauty, for the griddle of snowy mountains which surround the lovely valley, for its rivers and lakes and its primeval forests and for its moderated temperature. In prehistoric times the whole valley of Kashmir is supposed to have been covered by one enormous sheet of water, of which the Wular lake is the last remaining trace (Hassnain, 2002).

Apart from Wular lake the valley is bestowed with a number of magnificent water bodies. These water bodies are a home to the variety of flora, fauna and also to the fishing communities. It is not only the profession of this community that distinguishes them from the rest of the communities in the valley but also their housing pattern, food habits, ethnic wear and the jewellery is unique in itself. The women of the state of Jammu and Kashmir like their counterparts in other states also like to enjoy decorating themselves with the Jewellery. The traditional jewellery of Kashmir is specially made of silver. The sapphires mined in the region are also used to prepare jewellery in the state of Jammu and Kashmir. The Kashmiris are skillful artisans and artists, and their arts and crafts have been famous from ages (Hassnain, 2002). With the changing times the traditional jewellery has been replaced with the modern one. The heavy silver Jewellery is replaced with the gold jewellery. A long cloak called Firan with a distinct embroidery known as *Tilla* embroidery and traditional Jewellery worn by Kashmiri fisherwomen is unique in its style and beauty. In the words of W. R Lawerence, in his famous book "Valley of Kashmir", " Kashmir as possessing a distinct nationality, character, language, dress and body of customs affords much that is interesting, unique history and administration are worth careful study (Stein, 2005). Jewellery is an art in both design and workmanship. It therefore deserves to be studied as a major art form, on par with architecture, sculpture and paintaings (Chauhan, 2005). In the present study an attempt has been made to describe the ethnic Jewellery of Kashmiri fisherwomen under different categories, which are as follows:

1. **Head ornaments:** The fisherwomen of Kashmir wear a magnificent head gear studded with fancy needles and silver hangings. This head gear is known as *Kasab*. The *Kasab* consists of a cotton cloth around one meter long, a cap studded with fancy needles known as *Fali-tschin* and silver head band with hangings. The silver headband is known as *Tschini Poat* and the hangings as *Roni*. Though Kasab is not the favourite among the younger generation but the elderly fisherwomen still adore and wear it. The Kasab was a symbol of being married. On the occasion of marriages the bride's parents would call a lady locally known as *Kasbgarin* specialized in making *Kasab*. She would make the *Kasab* adding to the beauty of the bride

2. **Ornaments for neck:** Kashmiri fisherwomen wear necklaces with multifarious design, such as:

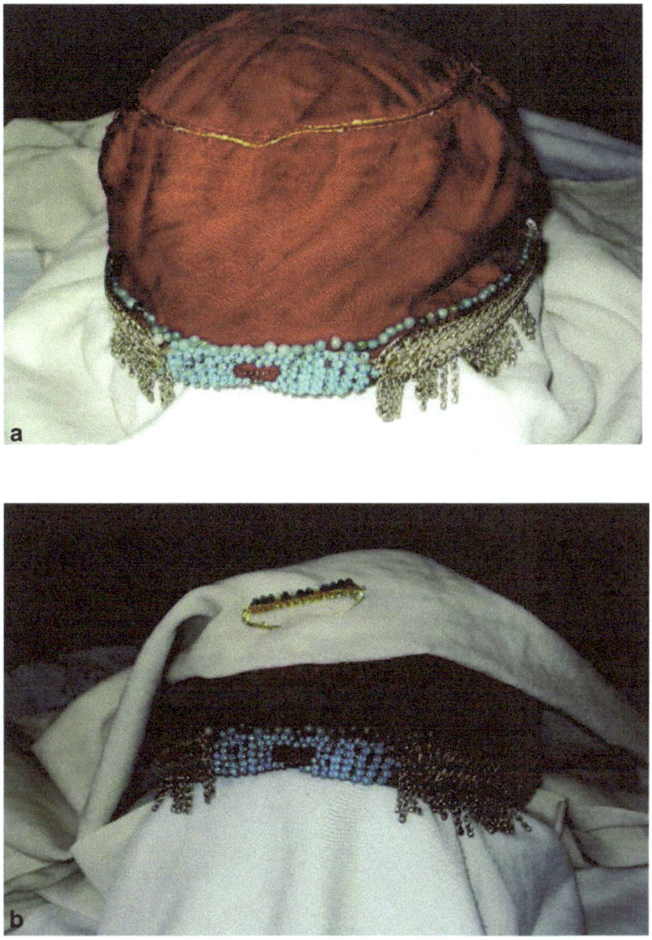

Fig. 1a,b: Ornaments for head

Fig. 1c: Fisherwoman wearing *Kasab* on head & *Firan* with tilla embroidery

Fig. 2a: Ornaments for neck

Fig. 2b: Ornaments for neck

Fig. 2c: Ornaments for neck

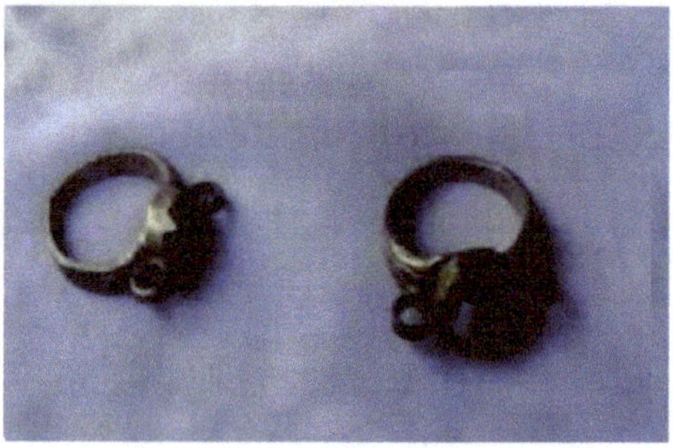

Fig. 3a: Ornaments for fingers

Fig. 3b: Ornaments for fingers

Fig. 3c: Ornaments for fingers

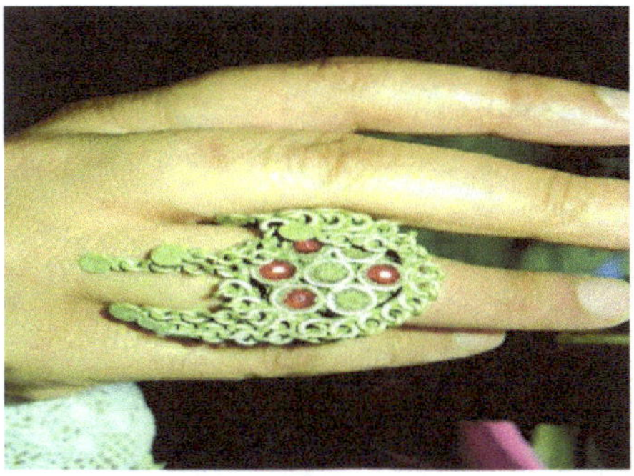

Fig. 3d: Ornaments for fingers

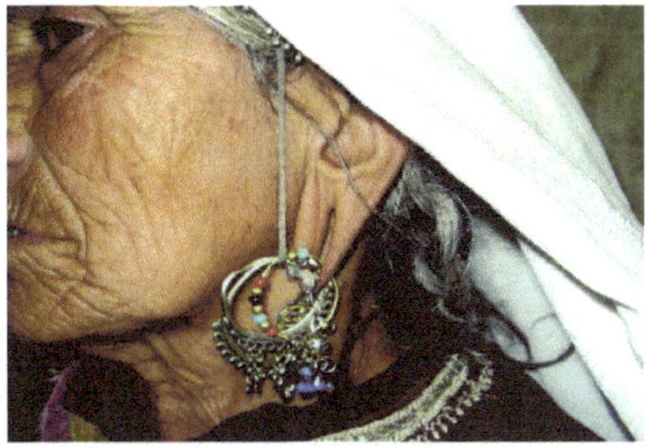

Fig. 4a: Ear ornaments

Fig. 4b: Ear ornaments

Fig. 5a: Ornaments for button holes

Fig. 5b: Ornaments for button holes

Fig. 6: Ornaments for Wrist

Fig. 7a: Ornaments for hair

Fig. 7b: Ornaments for hair

Fig. 8: Bridal chunni

a. *Round Beaded Necklace:* The necklace with round beads is made of silver with a heart shaped pendant in the centre. The pendant is locally known as *Bakwat*.

b. *Almond shaped beaded necklace:* The necklaces with the almond shaped beads are locally known as *Khail-maal*. The word *Khail* in Kashmiri language means almond. As the beads in the above necklaces are almond shaped, hence known as *Khail-maal*.

c. *Stone studded necklace:* The pretty stone studded silver necklace has a different design. Due to exclusive and vibrant design this necklace is liked and worn by the tourists while taking the pictures in traditional Kashmiri dress and jewellery.

3. **Ornaments for fingers:** The fisherwomen wear silver rings known as *Waj* in kashmiri. The finger rings of the fisherwomen have multifarious designs:

   a. *Simple finger rings:* These finger rings with split ends looks antique.

   b. *Studded rings with hangings:* These rings are studded with the stones and with the silver hangings on the split ends. The traditional trendy look of the finger rings definitely takes the people's looks away.

   c. *Studded rings:* These rings are usually studded with a single stone giving them a magnificent look.

4. **Ornaments for ears:** The silver ear rings worn by the fisherwoman are known as *Kan waji*. There are mainly two types of earrings used by the fisherwoman of Kashmir, which are:

   a. *Dangle earrings:* The dangle ear rings are favourite among the old fisherwomen. Some of these earrings have a mesh design whereas some have coloured beads. The fisherwomen wear multiple earrings in a single hole which give them a striking look.

   b. *Long earrings:* The lamp shaped long earrings have silver jinglings. The thin wire attachment in this ear ring penetrates the ear lobes easily. These are liked by the young female fisher folk.

5. **Ornaments for button holes:** The decorative fastener made of silver is fastened in the button holes of the neck of kurti. It is known as *Battan raz*. It consists of three short silver chains with connected ringlets forming a triangle shape at the two ends. It also has a silver button that is fastened in the button hole.

6. **Ornaments for wrist:**

   a. *Bangle:* A solid, cylinder bangle with a split worn by the fisherwomen is known as *Kor* in Kashmiri. It is made of silver and is aesthetically carved out giving it an antique look.

7. **Ornaments for hair:** The fisherwomen braid their hair with paranda with a silver jingling. The parnada with silver jingling is known as *Wank pan*.

8. **Bridal chunni with lamp shaped ear rings**: The duppata in the above picture is a bridal duppata. Ear ring like hangings are tied on the four corners of the duppata. The duppata is gifted by the bride's parents on the occasion of marriage.

References & Recommended Readings

Kak, R. C., 2002. Ancient monuments of Kashmir, Published by Gulshan Books, Srinagar, Kashmir.

Hassnain, F.M.K., 2002. Kashmir: The History of Himalayan Valley, Published by Gulshan Books, Srinagar, Kashmir, 67.

Stein, M. A., 2005. Ancient Geography of Kashmir, Published by Gulshan Books, Srinagar Kashmir.

Chauhan, H., 2005. The Traditional Jewellery of Pabbar valley. *Indian Journal of Traditional Knowledge* **4**(2): 118-126.

Chapter – 10

Documentation of Ethno-Entomophagy Practices in Ethnic Communities of Manipur, North East India

T. Shantibala, R.K. Lokeshwari, B.K. Agarwala and Gusheinzed Waikhom

INTRODUCTION

Ethno-entomophagy is the term used to describe the practice of consumption of insects as food by native human populations. Insects are an important source of protein for humans, either through direct consumption or as food supplements for livestock (DeFoliart, 1989). They are nutritious in terms of protein (40–75 g/100g dry weight) and minerals. Their protein is of high quality and has high digestibility (77–98%) with good concentration of essential amino acids (46–96% of the nutritional profile) (Verkerk et al., 2007). The biochemical composition and chemical analysis have revealed that they are comparable with other foods enriched with high proteins, fats, lipids, vitamins and mineral salts (Prasad, 2007). Insects generally have higher food conversion efficiency than other animals. They show a food conversion efficiency of about twice as efficient as pig and boiler chick and more than five times as efficient as beef cattle (Durst and Shono, 2010). Some of the important edible insects include grasshoppers, caterpillars, beetle grubs, adult winged termites, bees, wasps, winged ants, cicadas and a variety of aquatic insects. These insects hold the promise of exploitation as non-conventional food resources and represent a genuine food category

(Meyer-Rochow, 1976). Over 1500 species of edible insects have been recorded in 300 ethnic groups from 113 countries (MacEvilly, 2000). It is popular in countries like South Africa (Quin, 1959), Angola (Oliveira et al., 1976), China (Zhang et al., 2008), Zaire (Malaisse and Parent,1980), India (Meyer-Rochow and Changkija,1997), Thailand (Sungpuag and Puwastien, 1983), Mexico (Ramos-Elorduy, 1997), New Zealand, Korea, Japan (Mayer- Rochow, 2010), Australia and USA (Fromme,2005). Nevertheless, relevant information on this practice are scattered far and wide and there might be many more countries in the world that are yet to be explored for possible entomophagy by their populations.

Considering the popularity of the edible insects, it is not surprising that scores of species have been and are prominent items of commerce in the town and village markets of different regions of the world (De Foliart, 2002). The capturing, processing, transporting and marketing of edible forest insects provide important income and livelihood opportunities for an undetermined number of people. Traditionally, these activities were all locally based and largely under-recognized. Recently, more sophisticated and wide-reaching marketing or commercialization of edible insects has been advanced, including attractive packaging and advertising. It is also believed that creating a wider market for food insects could provide an economic incentive for conserving insect habitats (Johnson, 2010).

Manipur, being a distinctive part of the Indo-Burma biodiversity hotspot region, enjoys a very rich and fascinating diversity of insect fauna. The people of Manipur have been using insects in varied ways such as for edible, medicinal, industrial and cultural purposes. Of these, entomophagy is the most familiar practice. The state has 30 ethnic communities comprising 1 major group, the Meitei and 29 tribes with distinct identity, culture and food habit. Such a rich diversity of community has gifted the state with advantage for evolving innumerous knowledge on entomophagy. They have developed their own practice of eating insects through their traditional knowledge system. However, the modernization has led indigenous population around the world away from this traditional food source, without providing nutritional equivalent substitutes (Fromme, 2005). There is a serious threat for the erosion and even complete loss of this valuable knowledge forever unless certain effective measures are taken up. Therefore, the present paper aims to record the entomophagous fauna and document the different features of entomophagy practices prevalent in different ethnic communities of Manipur, India.

MATERIALS AND METHODS

Study Area

The state of Manipur lies in the North Eastern Region of Indian sub-continent, stretching between 23^{0}5′ N-25^{0}3′ N latitudes and 93^{0}4′ E-95^{0}3′ E longitudes (Fig. 1). It covers an area of 22,327 sq. Km, bordering Myanmar in the east, Nagaland in the north and Assam and Mizoram in the west. The region falls in the bio-geographic tri-junction of the Indian, the Himalayan and the Oriental landmass and forms a distinctive part of the Indo-Burma biodiversity hotspot. It is divided into nine administrative districts in which four districts namely Imphal East, Imphal West, Bishenpur and Thoubal are located in the central valley region and the rest five districts *viz.*, Tamenglong, Senapati, Chandel, Ukhrul and Churachandpur are situated in the hilly region.

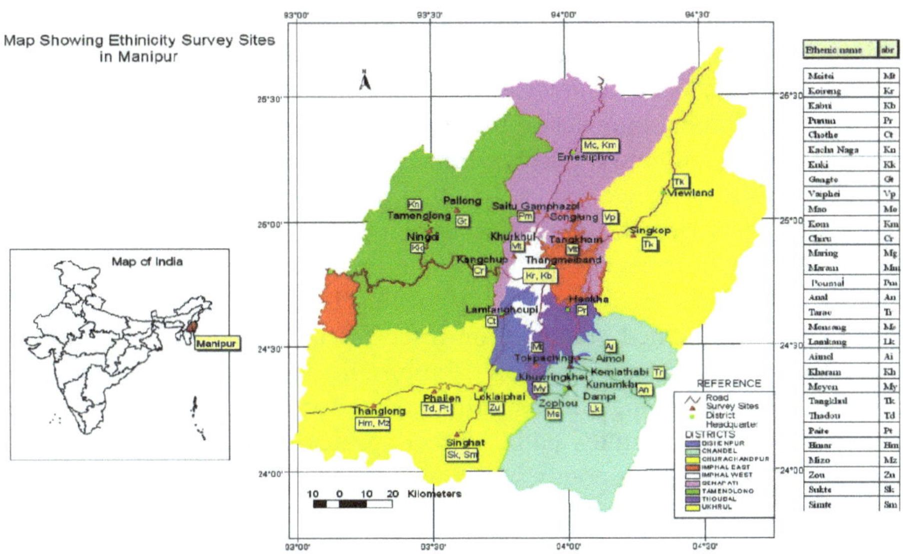

Fig. 1: Political Map of Manipur showing sites of study and name of ethnic groups distributed in nine districts of the state.

Ethnos

There are 30 recognized ethnic communities in Manipur consisting of 1 major group, the Meitei, and 29 tribes namely Aimol, Anal, Chiru, Chothe, Gangte, Hmar, Kabui, Kacha naga, Kharam, Koireng, Kom, Kuki, Lamkang, Mao, Maram, Maring, Mizo, Monsang, Moyon, Paite, Poumai, Purum Tarao, Simte, Sukte, Tangkhul, Thadou, Vaiphei and Zou (Devi,2006). The state has a population of 27.21 lakhs (2011 Census). Of the total population, 59.79% live in the valley and the remaining 40.18% are in the hill areas. The Meitei who live primarily in the state's valley region, form the primary ethnic group. The hilly region of the state is exclusively occupied by the tribes. However, these ethnic tribes are not evenly distributed. The main inhabited regions of these ethnic communities where the surveys were conducted are shown in the map (Fig.1).

Survey and documentation

Surveys for entomophagy practices in different ethnic groups were conducted in 25 villages inhabited by these ethnic groups (Fig.1) during January, 2008 to March, 2011. Data on different type of the insect species consumed by each ethnic group were obtained by performing personal interview from 105 persons comprising village heads, traditional knowledge holder, educated youth and homemakers. Information was collected from 3-4 targeted informants in each ethnic group. The age of these informants were above 20 years and included both man and woman. For the tribal ethnic community, each village has their own village head, locally known as the village king of that community. Village heads and traditional knowledge holders (TK holders) are aware about their age-old practices of entomophagy. Homemakers were included because they are responsible for collection, selling and preparation of the edible insects. Educated youths were taken to record their views about the habit of insect eating. Data were collected through a questionnaire attached with coloured photographs of the different insect species for easy identification. The questionnaire contained enquiries on insect species used as food, mode of consumption/utilization, form of preparations, life stages of insects consumed, association with other ingredients, culture related to insects and any other uses. During the course of study, edible insect species were also collected from different habitats such as ponds, lakes, streams, trees, shrubs, grassland, soils, paddy fields, dwellings, sericulture farms and forests. They were preserved following the standard methods (Ghosh and Sengupta, 1982) and identified up to family level by following the standard taxonomic keys (Castner, 2008). For the identification of the species, samples were sent out to institutes like United States Department of Agriculture (USDA) for hemipterans, Zoological Survey of India (ZSI), Chennai for Odonata, Indian Agricultural Research Institute for terrestrial coleopterans and ZSI, Pune for aquatic coleopterans etc. Voucher specimens and photographs were deposited in the Insect Bioresources Laboratory, Institute of Bioresources and Sustainable Development, Imphal, Manipur, India.

Use value

For each species of edible insect, use value (UV) was determined as UV= $\Sigma U/_{n}$, where, U is the number of citations per species and n is the number of informants (Philips et al., 1994). Use value is a functional parameter to evaluate the relative importance of the each species based on its use among informants. Such use value of each species is, therefore, based objectively on the importance attributed by the informants and does not depend on the opinion of the researcher.

Marketability of edible insects

Marketability of edible insects in the state was surveyed from seven major markets namely Kwairamband, Lamlong, Thoubal, Moirang, Tamenglong, Wino and Churchandpur bazaar situated in the districts of Imphal West, Imphal East, Thoubal, Bishenpur, Ukhrul and Churchandpur respectively. By adopting the purposive sampling among the marketers, only insect sellers (three insect sellers from each market) were interviewed (Osuala, 1986). Each market was visited three times during the survey period and informations on the species of edible insects offered for sale, units of measurement and cost per unit measure of the edible insects were recorded.

RESULTS

Diversity of edible insects

A total of 41 insect species belonging to 8 orders under 24 families and 34 genera (including 5 unidentified species of the order Hemiptera) are recorded as accepted food items (Table 1). This table also provides citation value, life stages of insects eaten and forms of consumption for each species of the edible insect. The species composition in different orders comprise of 1 species each in Dictyoptera and Isoptera, 4 species each in Lepidoptera and Orthoptera, 5 species in Coleoptera, 6 species in Odonata, 7 species in Hymenoptera, and 13 species in Hemiptera (Fig. 2).

The differences in the habit of consumption of insects among the 30 different ethnic communities are presented in Figure 3. It is found that five ethnic groups namely. Meitei, Tarao, Tangkhul, Chothe and Thadou consumed 28-30 species in comparison to 9-26 species consumed by other ethnic groups. Brief report of the species belonging to specific taxonomic orders viz., Odonata, Dictyoptera, Orthoptera, Isoptera, Hemiptera, Lepidoptera, Hymenoptera and Coleoptera are described hereunder.

Table 1: List of edible insects, their citation value, life stage eaten, form of consumption, preference of ethnic communities, seasonality and collection methods in Manipur

Order	Family	Scientific name	Citation value	Stage eaten	Mode of consumption*	Ethnic communities**	Seasonal availability	Mode of Collection method
Odonata	Libellulidae	*Pantala flavescens* (F.) (dragon fly)	59	nymph	B	1, 3, 4, 7, 8, 10, 13, 16, 17, 18, 19, 20,21, 22, 26, 27, 28, 29 = 18	Oct-July	dragging, scooping and noose gear
		Acisoma panorpoides Rambur (dragon fly) *Crocothemis servilia* (Drury) (dragon fly) *Orthetrum triangulare* (Selys) (dragon fly) *Rhyothemis variegata* (L.) (dragon fly) *Diplacodes trivialis* (Rambur) (dragon fly)						
Dictyoptera	Mantidae	*Heirodula* sp. (praying mantid)	44	nymph, adult	B	1, 2, 3, 4, 5, 6, 8, 11, 12, 13, 20,26, 27, 28, 29, 30 =16	Sept-Jan	hand picking/ netting
Orthoptera	Acrididae (rice grassho pper)	*Oxya hyla hyla* Serville	74	nymph, adult	B	1, 2, 3, 4, 5, 6, 7, 8, 9, 11, 12, 13,14, 15, 16, 17, 18, 19, 21, 22, 23, 24, 25, 26, 27, 28, 29, 30 = 28	Aug- Dec	hand picking/ netting
		Acridium melanocorne L. (short-horned grasshopper)	47	larva, adult	B	2, 3, 5, 6, 8, 9, 10, 11, 12, 17, 23,24, 25, 26, 27, 28, 29 = 17	June-Dec	hand picking/ netting
		Gryllus sp. (field cricket)	32	larva, adult	B	3, 6, 11, 14, 17, 20, 21, 24, 25, 27, 30 = 11	Oct-April	hand picking/ netting, light trap

(Contd.)

Order	Family	Scientific name	Citation value	Stage eaten	Mode of consumption*	Ethnic communities**	Seasonal availability	Mode of Collection method
	Gryllot alpidae	*Gryllotalpa orientalis* Burmeister (mole cricket)	48	larva, adult	B,D	1, 2, 3, 4, 5, 6, 7, 10, 11, 17,20, 22, 26, 27, 28, 29, 30 = 17	March -July	hand picking/ netting
Isoptera	Termitidae	*Odontotermes* sp.(termite)	50	queen, larva, adult	B	1, 2, 3, 7, 8, 12, 13, 15, 16, 17, 18,19, 22, 23, 28 = 15	Feb-Aug	hand picking/ netting
Hemiptera	Belosto-matidae	*Lethocerus indicus* (Lepeletier and Serville) (giant water bug)	53	nymph, adult	D	2, 4, 7, 8, 10, 13, 15, 16, 17, 18, 19, 20, 25, 27, 28 = 15	Nov-July	dragging, noose, entangle gear and light trap
		Diplonychus rusticus (F.) (water bug)	13	nymph, adult	D	17, 20 = 2	Sept-Dec	dragging, scooping and noose gear,
	Nepidae	*Ranatra* sp. (water scorpion bug)	16	nymph, adult	D	4, 17, 20, 28 = 4	June-July	dragging, scooping and noose gear,
		Laccotrephes maculatus Fabr.(water scorpion)	29	nymph, adult	D	1, 4, 17, 13, 18, 19, 20, 24 = 8		Sept-March dragging, scooping and noose gear,
	Dinidoridae	*Coridius* sp. (bug)	14	nymph, adult	C	9, 14, 18, 26 = 4	Oct-Dec	hand picking/ netting and scooping
	Pentat omidae	*Udonga montana* (Distant)(stink bug)	22	nymph, adult	B	2, 6, 18, 24, 25, 27 = 5	Nov-April	jerking
		Stink bug, unidentified species	20	nymph, adult	D	6, 16, 18, 21, 22, 26, 28, 29, 30 = 9	March-Aug	hand picking/ netting and jerking
		Stink bug, unidentified species	20	nymph, adult	D	6, 16, 18, 21, 22, 26, 28, 29, 30 = 9	April-July	hand picking/ netting and jerking
		Stink bug, unidentified species	20	nymph, adult	D	6,16, 18, 21, 22, 26, 28, 29, 30 = 9	April-July	hand picking/ netting and jerking
	Cicadidae	Cicada, unidentified species	20	nymph, adult	D	2, 3, 9, 16, 19, 26, 27 = 7	May-June	hand picking / netting

(Contd.)

Order	Family	Scientific name	Citation value	Stage eaten	Mode of consumption*	Ethnic communities**	Seasonal availability	Mode of Collection method
		Cicada, unidentified species	20	nymph, adult	D	2, 3, 9, 16, 19, 26, 27 = 7	May-June	hand picking / netting
	Corixidae	Water boatmen, unidentified species	11	adult	B	3, 4, 7, 17 = 4	Sept-Feb	dragging, scooping and noose gear,
	Noton	*Notonecta* sp (backswimmer) ectidae	11	adult	B	3, 4, 7, 17 = 4	Sept-Feb	dragging, scooping and noose gear,
	Hydro	*Hydrometra greeni* Kirkaldi	22	adult	D	4, 17, 20, 23, 26, 27, 28 = 7	Sept-Feb	metridae dragging, scooping and noose gear,
Lepidoptera	Bombycidae	*Bombyx mori* L. (silkworm)	14	Larva, pupa	A	4, 16, 22, 26, 27 = 5	March-Sept	hand picking
	Saturniidae	*Samia cynthia ricini* (Donovan)(eri-silkworm)	26	pupa	B,D	4, 6, 17, 16, 18, 21, 28 = 7	Perennial	hand picking
		Antheraea proylei Jolly (oak tasar silkworm)	10	pupa	A	8, 21, 26 = 3	March-Aug	hand picking
	Pyralidae	*Omphisa fuscidentalis* Hampson (bamboo worm)	81	larva	B	1, 2, 3, 4, 5, 6, 7, 8, 9, 10,11,12,14, 15, 16, 17, 18, 19, 21, 22, 23, 24, 25, 26, 27, 28, 29, 30 = 28	May-July	hand picking
Hymenoptera	Apidae	*Apis mellifera* L.(honey bee)	81	egg, larva, pupa	A,B	1, 2, 3, 4, 5, 6, 7, 8, 9, 10,11,12, 14, 15, 16, 17, 18, 19, 21, 22, 23, 24, 25, 26, 27, 28, 29, 30 = 28	Oct -Jan	smoking
		Apis cerana indica F. (honey bee)	81	egg, larva, pupa	A,B	1, 2, 3, 4, 5, 6, 7, 8, 9, 10, 11, 12, 14, 15, 16, 17, 18, 19, 21, 22, 23, 24, 25, 26, 27, 28, 29, 30 = 28	Oct-Jan	smoking

(Contd.)

Order	Family	Scientific name	Citation value	Stage eaten	Mode of consumption*	Ethnic communities**	Seasonal availability	Mode of Collection method
		Apis dorsata F. (giant honey bee)	81	egg, larva, pupa	A,B	1, 2, 3, 4, 5, 6, 7, 8, 9, 10,11,12, 14, 15, 16, 17, 18, 19, 21, 22, 23, 24, 25, 26, 27, 28, 29, 30 = 28	Oct-Jan	smoking
	Vespidae	*Vespa affinis* (L.) (social wasps)	78	egg, larva, pupa	A,B	1, 2, 3, 5, 6, 7, 8, 9, 10, 11,12,14, 15, 16, 18, 19, 21, 22,23, 24, 25, 26, 27, 28, 29, 30 = 28	Feb-May	smoking
		Polistes sp. (paper wasps)	80	egg, larva, pupa	A,B	1, 2, 4, 5, 6, 7, 8, 9, 10,11,12,14, 15, 16, 17, 18, 19, 21, 22, 23, 24, 25, 26, 27, 28, 29, 30 = 28	Oct-Dec	smoking
		Vespula vulgaris (L.) (yellow jacket wasps)	79	egg, larva, pupa	A,B	1, 2, 4, 5, 6, 7, 8, 9, 10, 11, 12,14, 15, 16, 17, 18, 19, 21, 22, 23, 24, 25, 26, 27, 28, 29, 30 = 28	April-Sept	smoking
	Formicidae	*Solenopsis geminata* F.(ant)	8	egg, larva, pupa	B	17, 27 = 2	Sept-Oct	hand picking
Coleoptera	Hydro philidae	*Hydrophilus olivaceous* F. (true water beetle)	78	grub, adult	B	1, 2, 4, 5, 6, 7, 8, 9, 10,11,12,14, 15, 16, 17, 18, 19, 21, 22, 23, 24, 25, 26, 27, 28, 29, 30 = 28	June-July	dragging, noose, entangle gear and light trap
	Dytiscidae	*Cybister confusus* Sharp (true water beetle)	78	grub, adult	B	1, 2, 4, 5, 6, 7, 8, 9, 10,11,12, 14, 15, 16, 17, 18, 19, 21,22, 23,	June-Feb	dragging, noose, entangle gear and light trap

(Contd.)

Order	Family	Scientific name	Citation value	Stage eaten	Mode of consumption*	Ethnic communities**	Seasonal availability	Mode of Collection method
						24, 25, 26, 27, 28, 29, 30 = 28		
	Curculi-onidae	*Cyrtotrachelus dux* Boheman (bamboo beetle)	73	grub, adult	B	2, 3, 4, 5, 6, 7, 8, 11, 12, 13, 14, 15,16, 18, 19, 21, 22, 23, 24, 25, 26, 27, 28, 29, 30 = 28	March-July	hand picking
	Dynastidae	*Oryctes rhinoceros* L. (rhino beetle)	38	adult	B,D	2, 3, 12, 13, 17, 18, 19, 20, 21, 27, 30 = 11	April-July	hand picking
	Ceramb-ycidae	*Anoplophora glabripennis* (Motschulsky)(long horned beetle)	36	grub	B	13, 14, 15, 16, 17, 19, 22, 27, 29 = 9	July-Sept	hand picking

*Mode of consumption: A- curry form; B-fried form; C-raw form; D-roasted form

**Ethnic communities: Aimol-1, Anal-2, Chiru-3, Chothe-4, Gangte-5, Hmar-6, Kabui-7, Kacha naga-8, Kharam-9, Koireng-10, Kom-11, Kuki-12, Lamkang-13, Mao-14, Maram-15, Maring-16, Meitei-17, Mizo-18, Monsang-19, Moyon-20, Paite-21, Poumai-22, Purum-23, Simte-24, Sukte-25, Tangkhul-26, Tarao-27, Thadou-28, Vaiphei-29, Zou-30.

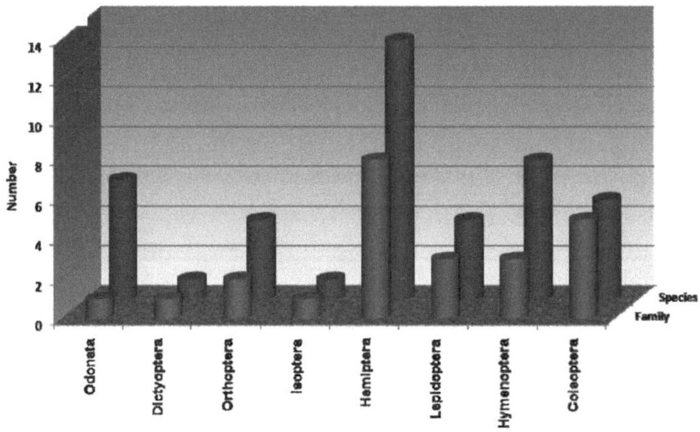

Fig. 2: Number of insect families and species under different groups eaten by different ethnic groups of Manipur.

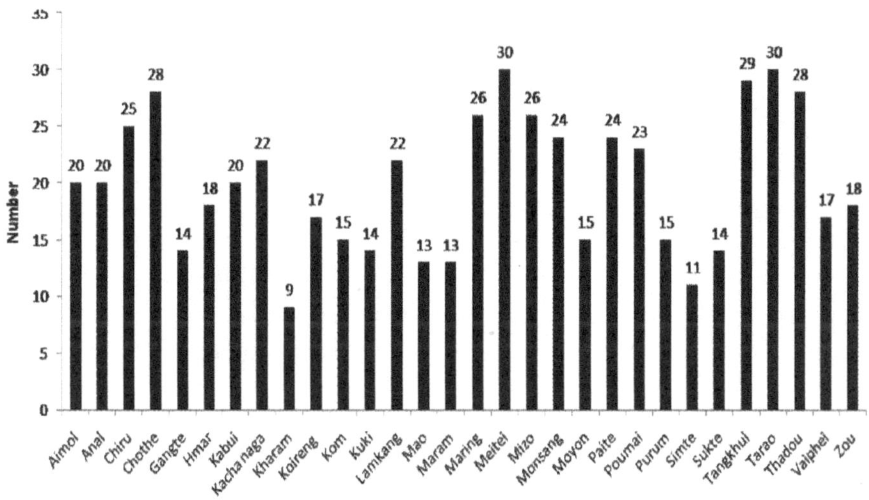

Fig. 3: Number of insect species eaten by different ethnic groups of Manipur.

Brief description of insect species in different orders

Odonata

Six species of dragonflies belonging to the family Lebullidae are recorded as common edible insects in the region. Nymphs are collected in clusters from flowing streams, lakes, drains and other wetlands by netting or a customized handmade device locally known as *Longdrai*. These are consumed in cluster without any preference for any particular species. Thus, all the six species under this order have similar citation value of 59 from 18 ethnic communities.

Orthoptera

In this order, four edible species of Orthoptera are identified under two families *viz.,* Acrididae (3 species) and Gryllotalpidae (1 species). One or more of these species are taken as food by all the 30 ethnic communities of the state. Most of the edible Orthoptera are the major crop pests and therefore, the consumption of these species is considered to be an effective means of their pest management. In Manipur and elsewhere in north east India, huge quantity of *Oxya hyla hyla* Serville from paddy field is collected during August to September for consumption by holding a large net at two ends and sweeping the crop.

Dictyoptera

One unidentified species of the genus *Heirodula* (Family: Mantidae) is consumed by people of 6 ethnic communities. Nymph and adult stage are harvested by people from vegetative stand and semi-forest areas. This species is regarded as an important delicacy and showed a citation value of 44. This insect was not reported earlier for entomophagy from the region.

Isoptera

Under this order one species of the genus *Odontotermes* (Family: Termitidae) was reported by informants of 15 ethnic communities as edible. Though, all the life stages are used for consumption, the most preferred stage is the large winged adult form (sexual form) due to its easy availability as compared to the other stages. Termites are usually harvested during swarming flight when they emerged from their nests for mating flights at the beginning of the rainy season during the months of February to August. A large swarm can be captured easily using mosquito net or hand picking. Further, other stages of the termite viz. eggs, workers, soldiers and queen are collected for consumption from termitarium as and when available.

Hemiptera

Thirteen species belonging to 8 families are recorded as the favourite food of informants of one or more of the 25 ethnic communities. Some of the important edible hemipterans are shown in Fig. 4. Giant water bug, *Lethocerus indicus* (Lepeletier & Serville) found to be the most preferred and delicious edible in the valley region. Their habitats comprise of shallow water of paddy fields, ponds and ditches. It has the highest citation value of 53. *Coridius* sp. is reported here for the first time from Manipur as edible insect among four ethnic communities. Another species, *Udonga montana* (Distant) is found to be utilized both for edible and extraction of edible oil. This insect is also a major pest of bamboo. Natural harvest of this insect in large quantity is done for oil extraction by the villagers during swarming which occurs during the flowering season of the bamboo (November to April) (Fig. 5).

Fig. 4: Examples of aquatic bugs as edible insects (a) *L. indicus* (b) *Coridius* sp.

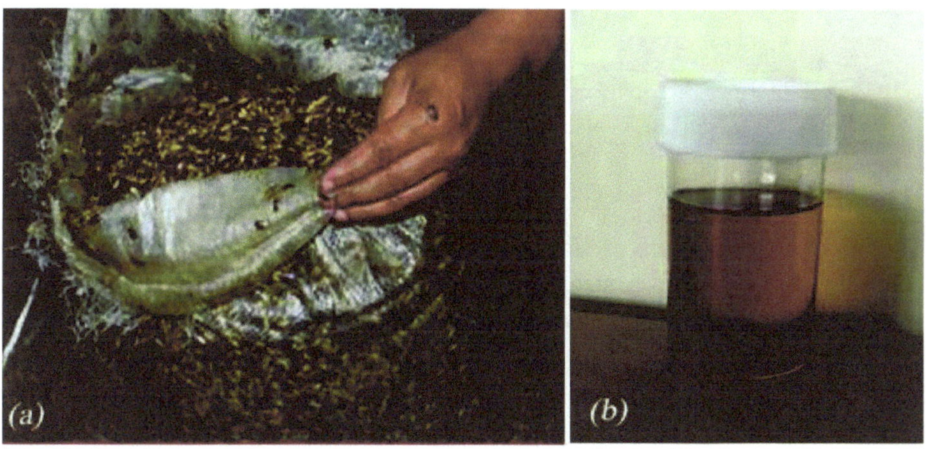

Fig. 5: Edible and oil yielding insect: (a) bag-ful stink bug, *U. montana* (b) extracted edible oil.

Lepidoptera

Four species in as many genera and 3 families are reported as edible insects in the present study. Larvae of *Omphisa fuscidentalis* Hampson (Family: Pyralidae) is the most common edible insect among the informants of 28 ethnic communities (citations = 81). It was noted that villager often cut tender shoots of the infested bamboos and collect the insect for edible purpose. A single infested node of bamboo yields, on average, 100-130 larvae (Fig. 6). Larvae and pupae of silkworms are also favourite as edible insects in certain ethnic communities. As Manipur is a prominent sericulture state, all the industrialized silkworm species can be easily available. In the meantime, some tribal peoples are also reared silkworm for household eating purpose.

Fig. 6: A colony of bamboo worm, *O. fuscidentalis*

Hymenoptera

Of these, bees and wasps are favorites in majority of the ethnic communities in Manipur. However, in contrast to the other species, *Solenopsis geminata* F. has got very less preference among the consumers. Seven species of hymenopterans belonging to three families are used as edible insects in Manipur.

Coleoptera

Five species belonging to as many families are recorded here as food. Two aquatic species, *Hydrophilus olivaceus* F. and *Cybister confusus* Sharp are found to be the most favourite in the state and were cited by 78 informants. *Crytotrachelus dux* Boheman is recorded here as another popular species which is consumed for its pleasant flavor. This species is a major pest of bamboos and causes enormous damages by tunneling inside soft tender tissue. Infested bamboos appear rotten and these are harvested for collection of the insect.

Insect collection methods/techniques

Insect collection methods/techniques are influence by various factors such as insect species, habited nature, characteristics of the raw material from which gear are fabricated. Insect collection methods range from catching with hands to the operation of indigenous nets. Beside this, still a number of the age old traditional methods are used. The techniques and gears used in fishing are also commonly utilized for capturing the aquatic edible insects. A single species can be harvested by adopting different methods. The techniques employed in insect collections are broadly categorized into eight different methods, which are as follows:

1. *Hand picking /Hand netting:* Generally, big size insects are collected simply by hands picking methods or using homemade hand net. This is the most common method adopted in collection of terrestrial insects. This practice requires deftness of hand and a considerable amount of practice. Insects like *C. dux, O. rhinoceros, A. glabripennis, O. fuscidentalis, G. orientalis, Heirodula* sp., *A. melanocorne, Odontotermes* sp. and *Cicada* sp. are caught by this method. Particularly for the ant species, *S. geminata*, precaution is to be taken up during collection. The whorl nest built in the plant is cut off and collected without disturbing the colony.

2. *Dragging gear method:* In this method, a medium sized (2 x1 m) net is towed through the water and drag till it reaches one bank of water bodies. At least two to four individuals are required to do the job. It is generally done in winter when the water level decreases. All the aquatic insects are commonly caught by this technique.

3. *Scooping gear method:* A scoop gear net locally known as 'longdrai' is made up of a nylon fibre tied in a rounded bamboo frame of about 1meter in diameter and ½ meter depth with a handle of about 5 feet long. Such net is usually operated in the shallow waters for capturing aquatic insects. During capturing process, the scoop gears is pushed inside the water bodies by holding at one end of the handle and shake vigorously then lift up above the water surface and the insects are collected. This technique is mainly used in the shallow water bodies like paddy fields, ditches, small channels etc.

4. *Noose gear method:* In this case, a long bamboo cane with a branching twig at one end is used for catching aquatic insects. The branching twig of the bamboo stick is pushed in-between the bushes of aquatic weeds

and twisted three-four turns to entangle them. Afterward, the whole bushes are dragged above the soil surface to collect the insects embedded with the aquatic weeds.

5. *Entangling gear method:* This type of gear is used to catch giant water bug. This gear used curtains of netting that are suspended by a system of floats and sinkers so that they can be anchored to the ground floor or allowed to float at the surface. The net is almost invisible to insects, so they swim right into it and accidentally entangled to their claws or legs. Entangled insects are then collected by lifting the net.
6. *Smoking method:* This method is practiced commonly in collection of insects belonging to the families Apidae and Vespidae. In general smoking is done during the night time at the foot of the trees or region where wasps or bee hives exist to sedate them so that the whole hive can be collected.
7. *Jerking method:* This method is employed in stink bugs collection. During the swarming period of these insects, the host plants are entirely covered with the same. The insect are collected in bulk by jerking the infested branches of the plants.
8. *Light trap method:* Under this method, nocturnal insect species that are attracted towards the light are captured. For instance, giant water bug, aquatic beetles, rhinoceros beetles etc which are attracted towards the household bulbs, street lamps and light houses are collected.

Use value

The use value of 41 different edible insect species observed in this study is presented in Figure 7. There is an asymmetry in their values even within the species belonging to same order and family. For instance *O. hyla hyla* because of its common occurrence in paddy fields, has greater use value (0.71) than *Acridium melanocorne* (0.45) of the same family, Acrididae. Among other species, *Apis dorsata* F., *A. mellifera* L., *A. cerana indica* (Family: Hymenoptera) and *Omphisa fuscidentalis* (Family: Lepidoptera) have the same highest use value of 0.78, due to delicacy of these species. *S. geminata* F., an ant species, has the lowest use value of 0.08 and its use is found to be limited to forest dwellers (Fig. 7).

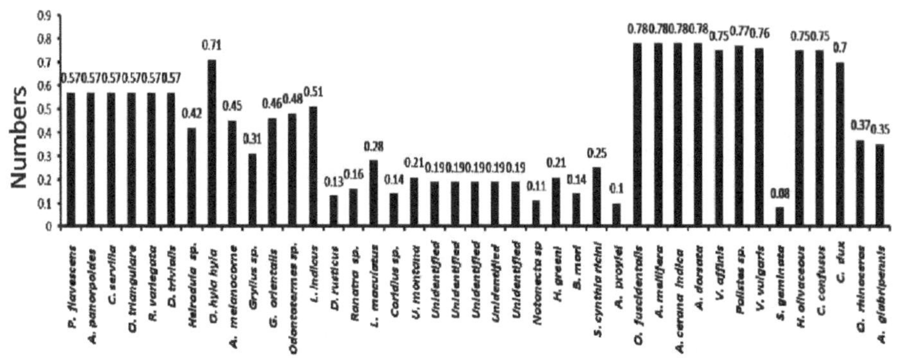

Fig. 7: Histograph showing use values of different edible insect species recorded in Manipur

Forms of preparation of edible insects

According to surveys, edible insects are prepared in any one of the four forms viz. curry, roasted, fried or raw (Table 1, Fig. 8). Depending upon the type of insects and form of the consumption, these are prepared with or without other ingredients like spices and oil. Hard bodied insects are eaten in roasted or fried form whereas soft bodied insects are eaten as curry or raw. The method of preparation is mostly traditional and is handed down from generation to generation. Nymphs/grubs and adults stages are mainly eaten in insects of Dictyoptera, Orthoptera, Hemiptera, Isoptera and Coleoptera. Odonatan insects are eaten in nymph stage only but lepidopterans are eaten in both larval and pupal stages. In case of hymenopterans, eggs are also eaten because most of these are social insects and produce eggs in large quantity. Methods of preparation as understood and recorded during surveys are described below.

i. Curry form: In this preparation, larvae and pupae of silkworms, honey bees and wasps are used. Soft bodied larvae and pupae are mixed with seasonings like oil, garlic, ginger, onion, salt and other spices according to individual's taste. Among the different sections of people, curry dish of these insects was treated as the most delicious item. Such dishes are especially popular for serving the special guests and are also liked by children. Aged members of the family are served with insect curry as a source of nutrition.

Insects are roasted as whole body, mixed with dried or fermented fish and then salted. Edible herbs are added to the preparation and a typical traditional dish called 'tin singju' is prepared. Dish is called 'kangshu' when crushed form of roasted insect is used.

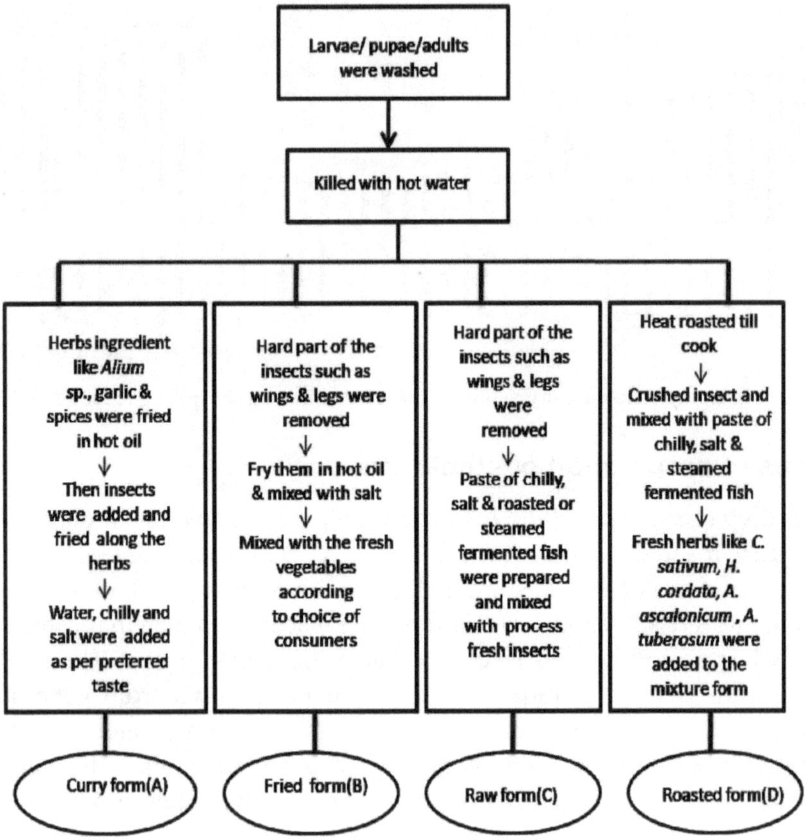

Fig. 8: Flowchart of preparation of different forms of edible insect (A) curry (B) fried (C) raw (D) roasted dishes.

ii. Fried form: This is the most common form recorded in Manipur in which 24 species of edible insects are eaten. Wings and legs of hard bodied insects are removed before cooking but soft bodied insects are fried as a whole. This practice is followed because wings and legs are highly sclerotized and are hard to digest. Fresh insects are fried with oil and then salted.

iii. Raw form: This form of consumption is found to be restricted to one particular edible species of the genus *Coridius* (Hemiptera: Dinidoridae) and is favourite of people in Ukhrul district of Manipur.

iv. Roasted form: Aquatic insects like giant water bug, aquatic beetles and water scorpion bug are eaten in this form.

MARKETABILITY

Edible insect species which are readily available and easy to capture are usually brought to markets for sale. The insect species sold in the market are presented in table 2. These include species of aquatic insects that are available throughout the year in ponds, lakes and ditches. Moreover, in large quantity these insects can be easily collected (Fig.9). Prices of the bees are usually high not only due to their taste but also for having cultural importance in tribal customs.

Fig. 9: A heap water beetles consisting of *H. olivaceous* and *C. confusus* for sale in a market

DISCUSSION

Insect species of diverse habits and habitats have been found to be eaten in different forms by different sections of the population in Manipur (Prasad, 2007) who listed 20 species of edible insects from this region. Brief accounts of the edible insects in north east India were recorded by Sachan et al. (1987) and Pathak and Rao (2000). Meyer-Rochow (2004) reported ten species of edible insects of the orders Odonata, Isoptera, Orthoptera, Hemiptera, Coleoptera and Hymenoptera, used by the Meitei community. However, these works did not reveal the extent of entomophagy prevalent in different ethnic communities which were the main objective of this study. The total numbers of the species recorded from Manipur are found to be higher than that recorded from the neigbhouring state. Nagaland (Ao and Singh, 2004) from where 15 insect species were reported as edible. However, 81 species of edible insect species have been enlisted from Arunachal Pradesh (Chakravorty et al., 2011). From other parts of India, reports on entomophagy are found in states like Tamil Nadu (Wilsanand, 2005), Orissa and Jharkhand (Srivastava et al., 2009). The list of

Table 2: List of edible insects available in the market of Manipur and their approx. price

Common name	Scientific name	Life stage	Approx. Price (USD)	Remark
Honey bee	*A. cerana indica*	larva, prepupa	4.17-10.42 per hive of 4x4 x2cm^3	This insect is favourite of Tangkhul ethnic community; serving of this insect curry to their married daughters or sisters is compulsory when they come to their parent home.
Honey bee	*A. dorsata*	larva, prepupa	10.42-16.67 per hive of 4x4 x2cm^3	This insect is very expensive because it is wild in nature.
Giant water bug	*L. indicus*	mature adult	0.35-0.52 per pieces of insect	Price of this insect varies depending upon its seasonal availability. Generally, price is high for male insects which produce aroma.
True water beetle	*H. olivaceous*	mature adult	0.63-0.83 per container (½ kg sized)	This insect is a common site in most markets.
True water beetle	*C. confusus*	mature adult	0.63-0.83 per container (½ kg sized)	This insect is sold without differentiation with *H. olivaceous.*
Dragon fly	Different dragon fly nymphs	nymph	1.04-2.08 per 1Kl	Dragon fly nymphs are sold as cluster of different species.
Bamboo worm	*C. dux*	larva	0.42 per 5 insects	Sold in clips of about 5 to 10 insects.

the edible insects reported from north-eastern states of India showed that the number of edible insect species is found to be higher in this region in comparison to the rest of India. Higher number of insect species used as food by the people of Meitei and Chothe, closely related tribes, could be due to their inhabitation in the periphery of the valley which has a high incidence of agricultural pests and aquatic insects. On the other hand, Thadou, Tangkhul and its descendent tribe, Tarao (Devi, 2006), are mainly forest dwellers and eat maximum numbers of forest insects like bees, wasps and bugs. Insects like rice grasshopper, bees, bamboo worm and aquatic beetles are commonly consumed by most of the ethnic groups. Ethnic-based practice of entomophagy have also been reported in Nyishi tribe (Singh et al., 2007), in Galo tribe (Kato and Gopi, 2009) and in a comparative study of Galo and Nyishi tribes of Arunachal Pradesh (Chakravorty et al., 2011). Yi et al. (2010) reported that the edible insects are the principal source of economy in 25 national minorities in Yunnan, Southwestern China. This suggests that the use and exploitation of the ecosystem varies according to the regions where humans settle and that people tend to use those species that are more readily available, more plentiful and easy to capture, store and prepare for eating (Ramos-Elorduy, 2009). Considering the importance of edible insects in the region, species conservation is today of primary interest because it means life. According to Posey (1990), indigenous people conserve 90% of the planet's germplasm. This highlights the appreciation that indigenous people have towards resources. Their primitive methods have permitted sustainable use of resources that leads to the survival of species (Ramos-Elorduy, 2009).

Results of this study showed that the nymph stage of dragon flies is used as food in ethnic communities of Manipur which is similar to entomophagy prevalent in ethnic communities of China and elsewhere in India (Chen et al., 2009; Singh et al., 2007). Three species of the family Lebullidae (order: Odonata), *Acisoma panorpoides* Rambur, *Crocothemis servilia* (Drury) and *Diplacodes trivialis* (Rambur) are found to be common with the species reported from Nagaland, Arunachal Pradesh and China. However, in case of Dictyoptera, more number of species is reported from Arunachal Pradesh (Singh et al., 2007) than the species reported in this study. Similar to the results of this study, orthopteran insects are among the most commonly reported edible in other asian countries like China (Chen and Feng, 1999) and Thailand (Hanboonsong, 2010). Earlier, three species from Manipur (Gope and Prasad, 1983), 10 species from Nagaland (Meyer-Rochow and Changkija, 1997) and 17 species from Arunachal Pradesh (Chakravorty, 2009) were reported as edible under this order. Termites, on the other hand, are also favorite food item in the state. Earlier reports on termites as human food was noticed from the south (Maxwell-Lefroy, 1909 and Rajan, 1987) and north eastern parts (Gope and Prasad, 1983) of India. The present report on the extraction and the use of edible oil from *U. montana* from Manipur is a rare phenomenon. Pathak and Rao (2000) described the methods of collection and oil extraction from this insect by Mizo tribes. Larvae of *O. fuscidentalis* (Lepidoptera: Pyralidae), which is reported as the most common edible insect among the informants of this study, was also cited as a notable edible insect in China (Chen et al., 2009) and Thailand (Hanboonsong, 2010). Interestingly, this insect is a serious pest of several bamboo-species, *Dendrocalamus* and *Bambusa* in particular (Chunram, 1993). Earlier study in China has reported a yield of 0.3-0.5 kilogram from one bamboo (Yi et al., 2010). Consumption of silkworm larvae and pupae was also found to be a common practice in Nagaland (Meyer-Rochow, 2004) and in China (Yi et al., 2010) and Africa (DeFoliart, 1995). This study revealed that bees and wasps are also prominent edible insects in Manipur. Earlier studies reported 15 species of hymenopteran insects as edible from Arunachal Pradesh (Chakravorty et al., 2011) and 32 species from China (Chen et al., 2009). During survey it was noted that people collect bee-hives and wasp-hives from forests or nursery and take out broods to be used as food by roasting, which is a similar practice of ethnic tribes of Mizoram (Pathak and Rao, 2000). People of the hill districts in Manipur considered the curry preparation of bees as the top most delicious item among their dishes. In other parts of northeast India, several hymenopteran insects are eaten throughout the year (Ao and Singh, 2004). Aquatic beetles are also popular in Manipur and other parts of north east India (Prasad, 2007). They are found throughout the year due to the presence of a large number of water bodies in the area of study and elsewhere in the region of Japan (Costa-Neto, 2002), China, Thailand, Indonesia and Vietnam (Yhoung-Aree and Viwatpanich, 2005).

Present studies on the mode of harvest of edible insect showed that different collection techniques was found to be employed depending on the prevailing situations of the insect's habit and habited at a particular period. The availability of edible insect is unpredictable, both in time and location (Banjo et al., 2006). Most edible insects are harvested from wild (DeFoliart, 1995). Same species of insect for example *C. confusus* was collected by using different methods like light trap, dragging gear, scooping gear, noose gear etc. Ramos-Elorduy et al. (2009) also described utilization of various devices and techniques in collection of aquatic insects by women, children or men using baskets, cloth nets or manually. For the collection of wasps and bees, smoking method is commonly practiced. Likewise, Nonaka (2010) depicted that when the entrance to the nest of *vespula* sp. is discovered, smoke is used to sedate the wasps inside. The nest is dugout while the wasps are sedated. The most common as well as primitive and age old technique for insect collection is the hand picking method. For instance, various species of grasshoppers and mole cricket (Chung, 2010) as well as *Rhynchophorous* sp. (Ramandey, 2007) are collected by this aged method. However, for some particular species like adult ants, sufficient care must be taken to create least disturbance while doing hand picking collection as well as to avoid from their bites and toxins (Adalle and Cervancia, 2010).

The preceding results suggested that diverse insect species are consumed in the state of Manipur. Selection of species as food is found to be dependent on their taste, favour and availability in the natural habitats (Ferreira et al., 2009 and Alves et al., 2009) and, therefore, different insects have different use. This result was also depicted by the use value of each species. Further, the way in which these insects are prepared into edible form is noteworthy. Preparation and cooking methods vary with the ethnic group (Sutton, 1995). Exopterygote insects (Orthoptera, Isoptera, and Hemiptera) are eaten in their nymph and adult stages whereas majority of the endopterygote insects are eaten in larval and pupal stages (Ao and Singh, 2004).

In Shillong, dish of grasshoppers is prepared by adding slice of onion, ginger, green chilies, salt, mixture of spices and condiments (Pathak and Rao, 2000). Fresh coriander, onion, cabbage etc are added to increase the taste and flavor according to the choice of consumers. In the state of Nagaland, most of the insects are dry fried after draining by boiling (Ao and Singh, 2004). On the other hand, pupae and larvae of bees and wasps are eaten as raw in some parts of north east India (Pathak and Rao, 2000) and queen termites are fed raw to weak children in parts of south India, especially in some villages of Karnataka state (Rajan, 1987). Similar forms of preparation were also described for giant water bugs (Prasad, 2007) and pentatomid bugs (Chakravorty et al., 2011) from Arunachal Pradesh. Some tribals of the region consumed giant water bugs in a different way in that they push dry rice inside the inside body and boiled it in water and then roasted (Pathak and Rao,2000).

It was interesting to note that many species of edible insects are found to be prominent as marketed items among the food commodities. There are reports of sale of grasshoppers and pine caterpillar @ USD 0.55-0.84/Kg and @ USD 0.84-1.02/Kg, respectively in different markets of north east India (Pathak and Rao, 2000) and in south India (Rajan, 1987). Johnson (2010) reported that in Asia and Pacific regions, public research has paid little attention to the subject of marketing and commercialization of edible forest insects. Absence of economic data represents a serious constraint to the commercial development of edible insects as commodity of trade between countries.

CONCLUSION

In the present study, 41 insect species are recorded as food insect by the people of different ethnic communities in Manipur, north east India. This observation indicated the significance acceptance of the insect as a respectable food item and also its role in promoting noticeable economic input. This useful biomass would be significant and comparable with other conventional food products by integrating scientific validation to the traditional wisdom through bio-prospecting. Documenting the significance of food insect becomes the foremost requirement in bringing linkage between people's livelihoods and assesses their sustainable use. Thus, further studies on insects as food should include key factors like taxonomy and ecology, management and conservation implications, industrialization and marketing of edible insects in order to promote their sustainable usage.

ACKNOWLEDGEMENTS

Authors are thankful to Department of Biotechnology (DBT), New Delhi for providing financial support and working facilities in the Insect Bioresources Laboratory, Institute of Bioresources and Sustainable Development. Thanks are also due to the United States Department of Agriculture, USA; Zoological Survey of India, Kolkata and Indian Agricultural Research Institute, New Delhi for the identification of the insect species. Authors are grateful to all the respondents for their cooperation in preparation of this document. The second author is also thankful to DBT for providing financial help through the DBT-Research Associateship grant under Biotechnology and Life Sciences program.

References & Recommended Readings

Adalla, C. B. and Cervancia C. R., 2010. Philippine edible insects: a new opportunity to bridge the protein gap of resource-poor families and to manage pests. In Proceedings of a workshop on Asia-Pacific resources and their potential for development: 19-21 February 2008, ed. Durst P. B., Johnson, D., Leslie, R. N., Shono, K., Thailand, pp. 151-164.

Alves, R. R. N., Barbosa, J. A. A., Santos, S. L. D. X., Souto, W. M. S. and Barboza, R. R. D., 2009. Animal-based remedies as complementary medicines in the semi-arid region of Northeastern Brazil, eCAM.

Ao, M. A. and Singh, H. K., 2004. Utilization of insect as human food in Nagaland. *Indian J. Entomol.* **66**: 308-310.

Benjo, A. D., Lawal, O. A. and Songonuga, E. A., 2006. The nutritional value of fourteen species of edible insects in Southwestern Nigeria. *Afr. J. Biotechnol.* **5**: 298-301.

Castner, J. L. 2008. Photographic Atlas of Entomology and Guide to Insect Identification. Gainesville, F.L., Feline Press, USA.

Chakravorty, J., Ghosh, S. and Meyer-Rochow, V. B., 2011. Practices of entomophagy and entomotherapy by members of the Nyishi and Galo tribes, two ethnic groups of the state of Arunachal Pradesh (North-East India). *J. Ethnobiol. Ethnomed.* **7**: 5.

Chen, X. M., Feng, Y., 1999. The edible insects of China. Science and Technology Publishing House, Beijing.

Chen, X., Feng, Y. and Chen, Z., 2009. Common edible insects and their utilization in China. *Entomological Research* **39**: 299-303.

Chung, A. Y. C., 2010. Edible insects and entomophagy in Borneo. In Proceedings of a workshop on Asia-Pacific resources and their potential for development: 19-21 February 2008, ed. Durst P.B., Johnson, D., Leslie, R.N., Shono, K., Thailand, pp. 141-149.

Chunram, S., 1993. Bamboo weevil. *Kasikorn Journal Bangkok* **66**: 483-484.

Costa-Neto, E. M., 2002. The use of insects in folk medicine in the state of Bahia, northeastern Brazil, with notes on insects reported elsewhere in Brazilian folk medicine. *Hum. Ecol.* **30**: 245-263.

DeFoliart, G. R., 1989. The human use of insects as food and animal feed. *Bull. Entomol. Soc. Am.* **35**: 22-35.

Defoliart, G. R., 1995. Edible insects as minilive stock. *Biodiversity and Conservation* **4**: 306-321.

De Foliart, G. R. 2002. The human use of insects as a food resource: a bibliographic account in progress. [http://www.food-insects.com/book.]

Devi, K. S., 2006. Introduction to Tribal Research Institute. Directorate of Development of Tribal and Scheduled Castes, Government of Manipur, Manipur.

Durst, P. B. and Shono, K.. 2010. Edible forest insects: exploring new horizons and traditional practices. in proceedings of a workshop on Asia-Pacific resources and their potential for development: 19-21 February 2008, ed. Durst, P. B., Johnson, D., Leslie, R. N. and Shono, K., Thailand, pp. 1-4.

Ferreira, F. S., Brito, S. V., Ribeiro, S. C., Almeida, W. O. and Alves, R. R.. 2009. Zootherapeutics utilized by residents of the community Poco Dantas, Crato-CE, Brazil. *J. Ethnobiol. Ethnomed.* **5**: 21

Fromme, A.. 2005. Edible insects. ZooGoer, 34. [http://nationalzoo.si.edu/Publications/ ZooGoer/ 2005/4/edibleinsects.cf]

Ghosh, A. K. and Sengupta, T. 1982. Insect collection, preservation and study. Zoological Survey of India, India.

Gope, B. and Prasad, B., 1983. Preliminary observations on the nutritional value of some edible insects of Manipur. *J. Adv. Zool.* **4**: 55–61.

Hanboonsong, Y., 2010. Edible insects and associated food habits in Thailand. In Proceedings of a workshop on Asia-Pacific resources and their potential for development: 19-21 February 2008, (ed.) Durst P. B., Johnson, D., Leslie, R. N., Shono, K., Thailand, pp. 173-182.

Johnson, D. V., 2010. The contribution of edible forest insects to human nutrition and to forest management. In Proceedings of a workshop on Asia-Pacific resources and their potential for development: 19-21 February 2008, ed. Durst, P. B., Johnson, D., Leslie, R. N., Shono, K., Thailand, pp. 5-22

Kato, D. and Gopi, G. V., 2009. Ethnozoology of Galo tribe with special reference to edible insects in Arunachal Predesh. *Indian J. Entomol.* **8**: 81-83.

MacEvilly, C., 2000. Bugs in the system. *Nutr. Bull.* **25**: 267-268.

Malaisse, F. and Parent, G., 1980. Les chenilles comestibles du Shaba meridionnale (Zaire). Les. Natual. *Belges*. **62**: 2–24.

Maxwell-Lefroy, H.,1909. Indian insect life: A manual of the insects of the plains (tropical India). Ed. Thacker, Spink and Co, Calcutta and Simla.

Mayer- Rochow, V. B., 2010. Entomophagy and its impact on world cultures: the need for a multidisciplinary approach. In Proceedings of a workshop on Asia-Pacific resources and their potential for development: 19-21 February 2008, ed. Durst, P. B., Johnson, D., Leslie, R. N., Shono, K., Thailand, pp. 23-36.

Merrit, W. R. and Cummins, K. W., 1996. An *introduction to the aquatic insects of North America.* Kendall/Hunt Publishing Co. Iowa, U.S.A.

Meyer-Rochow, V. B., 1976. The use of insects as human food. *Food Nutr.* **33:** 151-152.

Meyer-Rochow, V. B., 2004. Traditional food insects and spiders in several ethnic groups of Northeast India Papua New Guinea, Australia, and New Zealand. In Ecological implications of minilivestock: Rodents, frogs, snails, and insects for sustainable development, ed. Maurizio, G.P., U.S.A., 385-409.

Meyer-Rochow, V. B. and Changkija, S., 1997. Uses of insects as human food in Papua New Guinea, Australia, and North-East India: cross cultural considerations and cautious conclusions. *Ecol. Food Nutr.* **36**: 159-187.

Nonaka, K., 2010. Cultural and commercial roles of edible wasps in Japan. In Proceedings of a workshop on Asia-Pacific resources and their potential for development: 19-21 February 2008, ed. Durst, P. B., Johnson, D., Leslie, R. N., Shono, K., Thailand, pp. 123-130.

Oliveira, J. F. S., Carvalho, S. J., Passos de Sousa, R. F. X., Bruno de Simao, M., Sinao, M. M., 1976. The nutritional value of four species of insects consumed in Angola. *Ecol. Food. Nutr*. **5**: 91–97.

Osuala, E. C. 1986. Introduction to research methodology. Africana-FEP Publishers, Onitsha, Nigeria.

Pathak, K. A. and Rao, K. R., 2000. Insect pests as human food of tribals in North Eastern Region of India. *Indian J. Entomol.* **62**: 97-100.

Phillips, O., Gentry, A. H., Reynel, C., Wilki, P., Ga′vez-Durand, C. B., 1994. Quantitative ethnobotany and Amazonian conservation. *Conserv. Biol.* **8**: 225–248.

Posey, D. A., 1990. Introduction to ethnobiology: its implications and applications. In: Proceedings of the First International Congress of Ethnobiology on Ethnobiology: Implications and Applications, ed. Posey D. A., Overal, W. L., Museu Paraense Emílio Goeldi, Belém, pp. 1–8.

Prasad, B., 2007. Insect diversity and endemic bioresources. In Endemic Bio-Resources of India-Conservation & Sustainable Development with Special Reference to North East India, ed. Singh, N. I., Dehra Dun, 275-291.

Quin, P. J., 1959. Foods and Feeding Habits of Pedi. Johannesburg, Witwatersrand University Press, South Africa.

Rajan, B. K. C., 1987. Tiny wild fauna and human food. *My Forest* **23**: 177-180.

Ramandey, E., 2007. Debetkun Sebagai Alternatif Sumber protein. SUGAPA Edisi III. (January-March), 3-4.

Ramos-Elorduy, J., 2009. Anthropo-entomophagy: Cultures, evolution and sustainability. *Entomological Research* **39**: 271–288.

Ramos-Elorduy, J., Moreno, J. M. P., Camacho, V. H. M., 2009. Edible aquatic coleopteran of the world with an emphasis on Mexico. *J. Ethnobiol. Ethnomed.* **5**: 11.

Ramos-Elorduy, J., Moreno, J. M. P., Prado, E. E., Perez, M. A., Otero, J. L. and de Guevara, O. L., 1997. Nutritional value of edible insects from the State of Oaxaca, *Mexico. J. Food Compos. Anal.* **10**: 142-157.

Sachan, J. N., Das, B. B., Gangwar, S. K., Pathak, K. A. and Katiyar, J. N., 1987. Insects as human food in North Eastern Hill Region of India. *Bull. Ent.* **28**: 67-68.

Singh, O. T., Naobom, S. and Chakravorty, J., 2007. Edible insects of Nishi tribes of Arunachal Pradesh. *Hexapoda* **14**: 56-60.

Srivastava, S. K., Babu, N. and Pandey, H., 2009. Traditional insect bioprospecting- as human food and medicine. *Indian Journal of Traditional Knowledge* **8**: 485-494.

Sungpuag, P. and Puwastien, P., 1983. Nutritive value of unconventional protein source: insect. *Pochanagan Sarn.* **1**: 5–12.

Sutton, M. Q., 1995. Archaeological aspects of insect use. *J. Archaeol. Method Th.* **2**: 253-298.

Verkerk, M. C., Tramper, J., Van Trijp, J. C. M. and Martens, D. E., 2007. Insect cells for human food. *Biotechnol. Adv.* **25**: 198–202.

Wilsanand, V., 2005. Utilization of termite, *Odontotermes formosanus* by tribes of South India in medicine and food. *Natural Product Radiance* **4**: 121-125.

Yhoung-Aree, J. and Viwatpanich, K., 2005. Edible insects in the Laos PDR, Myanmar, Thailand, and Vietnam. In Ecological implications of minilivestock. Edited by Paoletti , M.G., Science Publishers Inc. Enfield, U.S.A,. 415–440.

Yi, C., He, Q., Wang, L. and Kuang, R., 2010. The utilization of insect-resources in Chinese rural area. *Journal of Agricultural Science* **2**: 146-154.

Zhang, C. X., Tang, X. D. and Cheng, J. A., 2008. The utilization and industrialization of insect resources in China. *Entomological Research* **38**: 38-47.

Chapter – 11

Traditionally Used Fish Feed Ingredients and Manure in Aquaculture in the North Eastern State of Tripura

M.K. Datta

INTRODUCTION

Over the past decades, as concerns about declines in local habitats, species, and livelihoods related to agriculture and allied sectors have increased, the potential contributions of local and traditional knowledge (LTK) to ecosystem research and management have been increasingly recognized (Thornton and Scheer, 2012). North Eastern region is well known as one of the 18 hotspot areas of the world and hence is blessed with rich biodiversity and fish fisheries resources. With more than 90% of population being fish eaters, there is heavy demand of fish but a wide gap exists between supply and demand. The region produces over 0.214 million tonnes of fish annually, with almost 50% coming from aquaculture.

Fish plays an important role in human nutrition and also contributes significantly to domestic food security of the nation (De Silva and Davy, 1992). In NE India particularly in Tripura, fish culture with 3 major Indian and 3 Chinese carps is carried out mostly in rural small ponds based on natural productivity and using manure and limited supplementary feed for higher yield. Empirical knowledge and experience gained over decades from these ponds had helped the farmers and workers to understand traditionally the methodologies of fish culture (Sinha, 1975).

Carp farming has attained commercial culture status in Tripura with participation of thousands of farmers for avocation though unit size of ponds remains very small (0.10ha). Both manuring for natural feed proliferation in pond and supplementary feeding is popularized for higher yield. Supplementary feeding offers best means of fish production within shortest possible time in ponds (Devaraj et al., 1986). The use of supplementary feed in carp culture has become inevitable for the success of fish culture (Shahzadi et al., 2006, Mohanty and Swamy, 1886).

SUPPLEMENTARY FEED USED IN AQUACULTURE PRACTICES OF TRIPURA AND FEEDING PRACTICES

Before independent, capture fisheries was dominating and only limited number of farmers was in aquaculture. Feed had very insignificant role in culture system. Use of cow dung, kitchen refuges and home produced rice bran were in use in ponds as fish feed. After independent, once the department of Fisheries established, gradually the aquaculture practices on scientific base initiated. As mostly composite carp farming was practiced by the farmers, the most popular combination of supplementary feed was advocated and adopted by farmers was Rice bran and Mustered oil cake in 50:50 ratio once to twice a day 2-5% of the total fish biomass. In the North-East mustard oil cake is generally used instead of ground nut oil cake due to its better availability. While some farmers only use rice bran, others use it in combination with 20-50% of oil cake. Most do not use vitamin and minerals. But due to lack of awareness and poor economic status, seldom the combination was in use. Among the other feed ingredients available in India, ground nut oil cake is most widely used feed ingredients in aquaculture but hardly used in NE India especially in Tripura. Sorghum and maize are used to a smaller extent than rice bran, principally as carbohydrate source for formulating commercial feed. Tapioca flour, which is rich in carbohydrates, is mainly used as binder. Among the few animal feed ingredients, fish meal is considered most important ingredient for aquaculture (De Silva, 1993). In commercial carp culture, rice bran and ground nut oil cake are most important feed ingredients nationally.

COMMONLY USED MANURE IN FISH POND

Manure is of organic nature and mostly used for enhancing soil fertility and natural productivity enhancement *i.e.* proliferation of palanktonice organisms in pond as fish natural feed. The most common organic manure used in pond is Raw cow dung @ 15,000-20,000 kg/ha/year traditionally due to their empirical knowledge about the importance of cow dung as manure in pond. Other organic matters used as manure is compost, chicken dung, pig dung, green leaves etc.

CONVENTIONALLY USED FEED AND FEEDING PRACTICES IN AQUACULTURE OF TRIPURA

Other than mustered oil cake and rice bran 1:1 ratio a fish feed combinations used by the farmers are rice dust (whole paddy is milled) 70%, MOC (20%) and dry fish (5%) weekly three times 5% of biomass in tray feeding. For poly culture of carp with prawn the popular combination is dust feed with MOC (1.2%), rice bran (1%), wheat bran (0.5%), dry fish dust (0.25%) and mineral mix (0.05%). Farmers also use the discarded breads etc. as fish feed.

TRADITIONAL LOCALLY AVAILABLE FEED INGREDIENTS AND MANURE USED IN TRIPURA

1. **Use of molasses:** Some farmers especially from tribal community used a combination of locally produced rice based molasses with cow dung/ with rice bran/with rice bran and MOC as supplementary feed to carps and reported good growth in fish as well as natural productivity. Use of molasses is reported in prawn raring and studied on the use of molasses as carbon source in limited discharge nursery and grow-out systems for *Litopenaeus vannamei* (Samocha et al., 2007). They suggested that molasses addition resulted in rapid reduction of TAN and nitrite-nitrogen and could be used to prevent buildup of these compounds and its application did not result any negative effect on water quality.

2. **Water hyacinth (*Eichhornia crassipes* Martius) as feed ingredient**: *Eichhornia crassipes* is an erect free-floating and stoloniferous perennial invasive herb, which belongs to the Family Pontederiaceae, available most of the natural wetland/ reservoirs etc. In Tripura, Water hyacinth is dried and dust is mixed with raw cow dung and mustered oil cake at a ratio of 20:50: 30 and used as feed to the carps in village ponds. The composted form of Water hyacinth is also used as pond manure in several ponds. Though it has negative impact of water chocking etc. But also has positive sides. Gopal (1987) reported that *E. crassipes* contains 64% methane and can be used for biogas generation and for water purification. It is also valuable in traditional medicine, making of ropes, production of fiber boards, as animal fodders and fish

feed, green manure, compost, and as an ornamental plant. *Eichhornia crassipes* was found to be rich in flavonoids, amino acids, crude protein, cyanide, phosphate, organic matter and organic carbon (Nyananyo et al., 2005).

3. **Duck weed as well as other aquatic plants as feed**: Duck weed like *Lemna, Azolla*, *Wolffia* etc. are used as feed particularly to grass carp and also to other carps in poly culture system from the naturally produced sources specially from unmanaged water bodies like beel/ wetland. Other aquatic plants used as fish feed are *Hydrilla* and *Vallisneria* as well as banana leaf particularly for grass carp. Large water bodies are mostly used as source of aquatic place growth and collected by the farmers on payment basis as feed to the fishes. The rate of one bag is around Rs. 50-100. Duck weed, a group of tiny, free-floating vascular plants with wide distribution are found in small water bodies such as fishponds, ditches and lakes, which are nutrient rich. Their ability to bloom within days after cultivation with high nutrient content has made them a rich source of food nutrients in the diet of fishes and animals alike. *Lemna* have been shown to exhibit antimicrobial activity (Skilicorn et al., 1993)

4. **Pineapple (*Ananas cosmosus*) and jackfruit (*Artocarpus heterophyllus*) seed**: As grow plenty in the state as summer cash crop, both seldom used as fish feed by the farmers. Jackfruit seed in meal condition as carbohydrate rich feed ingredient mixed with rice bran and mustered oil cake are in use. Seed of jackfruit at the rate of 10-15% are incorporated with other feed ingredients. The seeds give around 135 kcal/ 100 g. It is a rich source of complex carbohydrate, dietary fiber, vitamins like vitamin A, C and certain B vitamins, and minerals like calcium, zinc, and phosphorous. They contain lignans, isoflavones, saponins, that are called phytonutrients and their health benefits are wide-ranging from anti-cancer to antihypertensive, anti-ageing, antioxidant, anti-ulcer, etc. Pineapple both edible and mostly non edible after removal of spiny

portion used as feed by few farmers to the fishes. However, protein percentage is very low in the feed portion. 100 g fruit provides just about 50 calories. Its flesh contains no saturated fats or cholesterol; however, it is rich source of soluble and insoluble dietary fiber like pectin. Pineapple fruit contains a proteolytic enzyme bromelain that digests food by breaking down protein. Bromelain also has anti-inflammatory, anti-clotting and anti-cancer properties.

5. **Use of *Refugee Lata (Mikania cordata)* as feed and manure:** Many of the tribal farmers of the state are using this plant as fish feed as well as manure traditionally. It was introduced into India during World War II to camouflage air fields or as ground cover for tea plantations. It has spread to moist tropical and subtropical regions and the north-east of India and known as invasion of alien plant species is native to Central and South America. It has been reported that some biochemical substances obtained from this weed exhibit antibacterial and antimicrobial activity, and show inhibitory action against parainl uenza type 3 and respiratory syncytial viruses. Scientific study is needed about its utility as manure cum fish feed.

6. **Silk worm pupae (SWP):** Farmers of Tripura seldom use silkworm pupae as feed to the cultured fishes. Sericulture is a well-established rural agribusiness in India and produces about 40,000 MT per year of silkworm pupae (SWP) on dry weight basis. A small quantity of this SWP is sun-dried and used in animal feeds, while a considerable amount is discarded in open places. Disposal and sun-drying of SWP causes environmental pollution besides loss of nutrients. Digestibility of the crude protein in silkworm meal was found to be similar to fish meal (91.58 vs 89.68) when fed to fish. Silkworm meal can be used to completely replace fish meal in carp diets (Nandesha, 1995, Rangacharyulu et al., 2003). The crude pro-

tein content of silkworm pupae is around 54.2 to 65.5 in raw form. In Tripura, around 4,500 farmers are directly involved in the sericulture. As Tripura is one of the major producer of mulberry silk and production of silk worm residue in the processing centre is about 30 MT per/year (60-65% of silk worm left as cocoon waste) in wet weight basis, can be utilized as ingredient of regular fish feed.

7. **Sweet potato (*Ipomoea batatas*,** L. Lam.): Sweet potato is mainly grown as food for human consumption in Tripura extensively. The tubers used as feed to fish as a source of carbohydrates with a high content of sucrose. Protein content, although low, is almost twice that of cassava. The leaves of the sweet potato are highly digestible, free from toxins, and fairly rich in protein. The starch in the tuber provides good binding for dry and moist-type pellets if heat processing is employed.

8. **Cassava, tapioca or yucca (*Mannihot esculenta*, Crantz**); Fresh cassava carbohydrate rich feed is used mostly to pig and limited way to fish. However, most varieties are toxic unless properly processed to prevent the release of hydrocyanic acid by an enzyme present in the tuber. Heat applied during processing destroys the enzyme. Cassava has low feed value in diets for fish but its presence in small quantities improves the water stability of pelleted feed. Cassava is cultivated in the hills of Tripura.

9. **Kack-dumur (*Ficus hispida* L.) as fish feed:** Seed is reported to be used as feed to the fishes specially to prawn and. The seed and fruit are traditionally used for curing diabetes, dermatitis, emetic. It is locally known as *Dumoor* (Bengali), *Gobla* (Hindi) and *Peyatti* (Tamil) is a shrub or

moderate sized tree, grows well in damp locality and in shady places. Almost all parts of this plant are used in Indian traditional medicine for the treatment of various ailments. Bark as emetic and anti-diarrhoeal. Fruits as tonic, lactogogue, cooling, astringent and in hepatic obstruction. The plant was found to be used widely as an anti-diarrhoeal, hepatoprotective, antitussive, antipyretic, anti-inflammatory, depurative, vulnerary, haemostatic, anti-ulcer and in treatment of anaemia. From proximate analysis (mg/100 g) of leaf sample, total soluble sugar was estimated 50, phenols 22.36, tannin 21.8 and calcium 1.152.

10. **Awra (fuel wood semi wild plant) as fish feed**: Carp like specially rohu like rotten bark of awra plant (wild plant of tropic of NE India mainly used for firewood purpose) as feed. The stems of awra plants are placed inside the pond in different places and after about one week the barks become soft and browser like rohu like the bark as feed for their growth.

11. **Bamboo as medium of fish feed**: System of periphyton-based aquaculture was considered as an alternative to fish production in conventional, substrate-free, pond. In Tripura, bamboo is used extensively in pond as substrate as well as for avoiding pouching. Azim et al. (2003) demonstrated that fish production in the periphyton-based systems was higher than in the substrate-free controls. Commonly used bamboo in pond as substrate is *Melocana bambusoides* (common name: *Muli*) as well as *Bambusa polymorpha* (common name: *Bari*).

12. **Traditional herbs used as fish attractant used in feeding/ harvesting and angling:** Fish feeding pattern, whereby preferred foods are searched for and ingested, is one of the most important behavior pattern exhibited by fish and several herb species obtained from Tripura were studied with carps and prawn with positive result (Harpaz, 1997, Venkateshwarlu et al., 2009). Ekangi, tambul, kharboj, chotokakla and latakasturi are the plant materials that have been used for alluring fish during harvest and angling by traditional fish farmers. In Tripura several herbs are traditionally used as feed attractant especially for feeding, harvesting and angling of carps and air breathing fishes. The herbs either

rhizome, seed or roots, fruits are incorporated with conventional feed ingredients like MOC Rice Bran @ 1-2 %.These are collected locally and necessary processing is undertaken traditionally. Most of them have aromatic properties.

Local name	Plant name	Parts used
Ekangi	*Kaempferia galanga*	Rhizomes
Tambul	*Zanthoxylum acanthopodium*	Seeds
Kharboj	*Cucumis melo*	Roots
Chotokakla	*Piper cubeba*	Fruits
Latakasturi	*Hibisscus abelmoschus*	Seeds
Jatamansi	*Nordostachys Jatamansi*	Roots
Jayatri	*Myristica fragnans*	Flowersd
Kakla	*Piper cubeba*	Fruits
Latkhandhana	*Bixa orellana*	

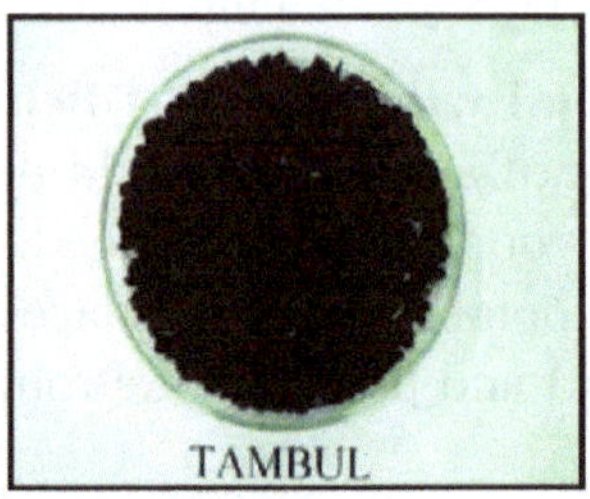

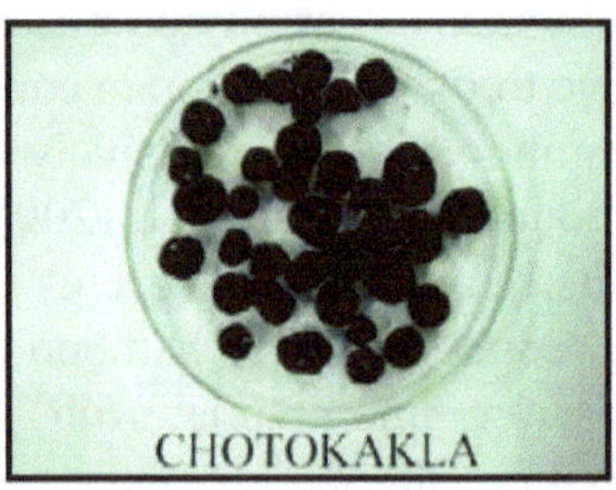

Source: Paul (2013)

13. **Traditionally used rice bran: boiled and un-boiled:** Rice bran is a major feed ingredient which is the cheapest among the entire feed ingredient used, commonly available and well accepted by the carps for all types fish farming in NE India. The definition of rice bran suggested by FAO (1996) is "a by-product of polishing brown rice, comprising the pericarp, aleurone layer, embryo and some endosperm". It is an underutilized by-product of the rice milling process. In commercial rice production, brown rice is commonly processed by polishing or rubbing the rice kernals

four times. This process produces rice bran among other by-product. Generally two types of rice bran are available in the market; boiled and un-boiled. Mostly in north east (NE) India rice bran used and commercially available in aquaculture is of boiled. Whereas un-boiled rice bran is available in limited pocket of location in NE India where rice is consumed and available in market in un-boiled condition. The two forms of rice bran contained different protein percentage (in un boiled rice bran 9.31 and in boiled rice bran 6.53). Rice bran is commonly used as energy supplement in carp culture in Asia. Rice bran as a single feedstuff has proven to be more economical than commercial fish feeds (Middendrop and Verreth, 1991, Mims et al., 1993).

Freshly produced rice bran has high oil content (14-18 percent). The oil is sometimes recovered by solvent extraction because of its high commercial value, especially in countries where demand for cooking oil exceeds supply. The oil present in the bran is rich in polyunsaturated fatty acids which undergo rapid oxidation under normal storage conditions. Rice bran that has turned rancid has markedly reduced feed value. Rice bran has higher protein content than the grain. It is also fairly high in fibre, thus limiting its use in fish feeds. Often, rice bran is mixed with rice polishing at the mill and sold as rice bran. Adulteration with hulls lowers its feeding value considerably. Good quality pellets can be made from feeds containing un-extracted rice bran

14. **Broken rice:** As such, the bulk of the ingredients used as supplementary feed are agricultural by-products (De Silva and Anderson, 1995). Rice (*Oryza sativa*) is the staple food for 2/3rd of the world population. Worldwide, more than 500 million MT rice is produced annually of which 92% is harvested in Asia. Rice is India's predominant crop, and is the staple food of the people of the eastern part of the country and particularly in Tripura.

    Whole rice hardly fed to fish. Instead, broken rice (polished) and unpolished rice are used in the preparation of traditional type fish feeds as well as in compound diet formulations. The rice is first boiled in water before mixing with other ingredients to make the ration. Rice is an excellent energy source with low but fairly good quality protein content. Uncooked rice is more digestible than maize. Unpolished rice, especially, is rich in the B-vitamins.

    Oil extracted rice bran is almost totally devoid of oil. It is the type of rice bran favoured by feed mills because of its longer shelf-life. Its protein content is somewhat higher than fresh rice bran while its energy content is lower. Pelleted feed containing a high proportion of de-oiled rice bran are not very water-stable. This is because of the water absorption characteristics of fiber, which will be fairly high in such feed. Where available, rice bran has been successfully used in compound fish feeds for both warm as well as cold water species (Garcia and Toledo, 1989).

15. **Rice polishing:** Rice polishing is produced during the scouring of rice after the bran has been removed. The material, therefore, contains a higher proportion of starch but is lower in protein and fiber content (Houston, 1972). It is not produced in as large quantities as rice bran, and, in most mills, are often mixed with the latter to give rice pollard. This, however, is usually sold as a grade of rice bran in most countries.

COMPOSITION OF FISH FEED INGREDIENTS

The important fish feed ingredients and their nutrient values are given below:

Ingredient	Average composition (%)						
Plant ingredients	H_2O	CP	EE	CF	Ash	Ca	P
Mustered oil cake	8.47	37.25	10.73	-	6.99	-	-
Ground nut oil cake	8.60	42.50	6.10	8.50	6.20	0.20	0.62
Rice bran	9.15	6.97	13.07	7.20	14.79	-	-
Jackfruit seed flour	14.07	9.03	1.10	2.55	2.55	-	-
Pineapple	86.00	0.54	0.12	1.40	0.22	-	-
Duck weeds	-	6.80	1.80	5.70	12.00	-	-
Azolla meal	-	21.40	2.70	12.7	16.20	1.16	1.29
Water hyacinth	-	11.87	1.18	21.79	15.88		
Hydrilla	-	17.10	2.79	13.34	19.45	1.29	
Soybean meal	9.30	40.80	5.70	5.30	6.30	0.22	0.56
Wheat grain	12.60	10.30	1.40	2.20	2.20	0.07	0.39
Wheat bran	11.60	12.80	3.10	8.60	4.50	0.10	0.89
Wheat flour	12.80	10.60	1.20	0.40	0.80	0.04	0.15
Animal ingredients							
Blood meal	10.80	73.50	0.40	-	5.80	0.14	0.06
Fishmeal	10.45	61.32	5.58	-	-	2.38	2.27
Silk worm pupae	61.30	25.50	9.70		1.60	-	-
Chicken viscera	78.10	5.70	11.40	-	1.40	-	-

CP= Crude Protein, EE= Ethyl Ester, CF= Crude Fiber
Source: Winter School Training Manual, CIFA, 2013

CONCLUSION

The traditional knowledge is empirical in nature needs nurture care and many of cultivation management practices as knowledge lost entity forever due to time and change in human anthropological activities. New scientific enrichment in modern farming methods and also in aquaculture more market demand driven. However, for cost effective sustainable management practices specially the high cost feed and manure sector, more emphasis is needed to save; scientifically validate the traditional practices and knowledge with scientific fortitude.

References & Recommended Readings

Azim, M. E., Milstein, A., Wahab, M. A., Verdegem, M. C. J., 2003. Periphyton water quality relationship in fertilized fish ponds with artificial substrates. *Aquaculture* **228**: 169-187.

De Silva, S. S., 1993. Supplementary feeding in semi-intensive aquaculture systems. In M. B. New, A. G. J. Tacon and I. Csavas (eds). Farm-made aquafeeds, p. 24-60. Proceedings of the FAO/AADCP Regional Expert Consultation on Farm-Made Aquafeeds. Bangkok, FAO-RAPA/AADCP.

De Silva, S. S. and Anderson, T. A., 1995. *Fish nutrition in aquaculture*. Chapman and Hall, N.Y., USA. pp.319.

De Silva, S. S. and Davy, F. B., 1992. Fish nutrition research for semi-intensive culture systems in Asia. *Asian Fish. Sci.* **5**:129-144.

Devaraj, K.V., Keshavappa, G. Y. and. Manissery, J. K., 1986. Growth of grass carp, *Ctenopharyngodon idella* (Val.), fed on two terrestrial fodder plants. *Aquaculture Fish Manag.* **17**: 123-128.

FAO, 1996. Draft Definition and Classification of Commodities, W2979.

Garcia, M. C. and Toledo, J., 1989. Effect of rice bran on growth of channel catfish fingerlings. (*Ictalurus punctatus*). *Rev. Latin American Agriculture* **39**:10-13.

Gopal, B., 1987. Water Hyacinth. Aquatic mat, Studies Series, ELSEVER, p74.

Harpaz, S., 1997. Enrichment of growth in fresh water prawn, *Macrobrachium rosenbergii* through use of a chemo-attractant, *Aquaculture* **156**: 221

Houston, D. F., 1972. Rice bran and polish. In: Houston, D.F.E. (Ed.), Rice: Chemistry and Technology. AACC Publications, Minnesota.

Middendrop, H. A. J. and Verreth, J. A. J., 1991. The development of small-scale hapa culture of tilapia (*Oreochromis niloticus*) in Northeastern Thailand. 2. The feasibility using low cost compound feeds. *Asian Fishery Science* **4**:317-327.

Mims, S. D., Clark, J. A.,Williams, J. C. and Rose, D. B., 1993. Comparison of two by-products and a prepared diet as organic fertilizers on growth and survival of larval paddle fish, *Polyodon spathula*, in earthen ponds. *Journal of Applied Aquaculture* **2**:171-187.

Mohanty S. N. and Swamy, D. N., 1886. Enriched conventional feed for India major carps. Network for aquaculture countries in Asia, Bangkok, Thailand, NACA/WP/86/28.

Nandeesha, M. C., 1995. Aquafeed and feeding strategies in India, Farm Made Aquafeeds edited. Michael, B. New; Albert, G. S., Tacon, G. K. and Imrc, Csavan, Fisheries Technical paper, FAO, 343.

Nyananyo, B. L., Ekeke, C. and Mensah, S. I., 2005. The Morphology and Phytochemistry of Water Hyacinth. *Journal of Creativity and Scientific Studies* **1**(2 & 3): 20-29.

Paul, B. N., 2013. Manual on sustainable fish feeds and neutraceutacal to grow health propmoting fish. Pub: CIFA. 165pp.

Pekar, F. and Olah, J., 1990. Organic fertilization. *In:* Proceeding of FAO-EI FAC symposium on production enhancement in still water pond culture, (eds. Berka, R. and Hilge, H.). Prague. Czechoslovakia: 116-122.

Rangacharyulu, P. V., Giri, S.S.; Paul, B.N., Yashoda, K.P., Rao, R. J., Mahendrakar, N. S., Mohanty, S. N. and Mukhopadhyay, P. K., 2003. Utilization of fermented silkworm pupae silage in feed for carps. *Bioresour Technol.* **86**(1):29-32.

Samocha, T. M., Patnaik, S., Speed, M. Ali., Josh, M., Burger, A. M., Almeida, R. V., Ayub, Z., Harisanto, M., Horowitz, A. and Brock, D. L., 2007. Use of molasses as carbon source in limited discharge nursery and grow-out systems for *Litopenaeus vannamei*. *Aquaculture Engineering* **36**(2): 184-191.

Shahzadi T., Salim, M., Kalsoom, U. and Shahzad, K., 2006. Growth performance and feed conversion ratio (FCR) of hybrid fingerlings (*Catla catla* X *Labeo rohita*) fed on cottonseed meal, sunflower meal and bone meal. *Pakistan Veterinary Journal* **26**(4): 163-166.

Skilicorn P., Spirar, W. and Journey, W., 1993. Duckweed Aquaculture: A new Aquatic farming system for Developing countries. A World Bank Publication 76P. National Agricultural Research Project (NARP), Nigeria. 309

Sinha, V. R. P. 1975. New trends in fish farm management. FIR: AQ/ Conf. 76/R.10, FAO Technical Conference on Aquaculture, Kyoto, Japan.

Thornton, T. F. and Maciejewski Scheer, A., 2012. Collaborative engagement of local and traditional knowledge and science in marine environments: a review. *Ecology and Society* **17**(3): 8.

Venkatateshwarlu, G., Muralidharan, A. P., Rathod, R. and Pal, A. K., 2009. Plants traditionally sed in fish harvest and angling potential feed attachments in aquaculture. *Indian Journal of Traditional Knowledge* **8**(4): 539-542.

Chapter – 12

Identification of Indigenous Resource and its Managements for Livelihood by Tribal Women Fishers of Dooars Region of West Bengal Using PRA Technique: A Case Study

Ambalika Ghosh

INTRODUCTION

Fishing is one of the oldest professions of humans. As the day went by, the fish farmers carry out their own indigenous technology as a common practice through natural resource management. In this line, different traditional fishing communities of West Bengal gained a vast experience in every sphere of aquaculture. Their innovation is based on their own knowledge derived from their day to day effort of solving problems. Actually this indigenous knowledge is the accumulated information, skill and technology of the local farmers derived from the close interaction with the nature. The intimate interaction of the farmers with aquatic ecosystems through their age old practices has developed huge expertise which is essential to be documented and disseminated. It will be helpful for aqua resource management as they can be shared in a, cost effective, eco-friendly as well as socio economically viable proposition.

"Participatory approaches" are used to describe a wide range of development and research approaches, methods and tools that are increasingly proposed as a means of improving the practice of development now a days many organizations and agencies adopting these approaches and attempting to apply them throughout their development programmes. Practitioners working in technical fields, such as fisheries and aquaculture are essential to carry out their work in a "participatory" way. A participatory approach means, a process in which we deeply respects the traditional wisdom of the local people and creates an environment in which local people can understand their potential, their capacity, and their power and come to develop their self esteem.

PRA technique has been utilized by many in India and abroad, but to my knowledge, no report is available on application of PRA technique for resource mapping. Therefore in this study Participatory Rural Appraisal has been utilized in some selected villages surrounding the Rava Basti and Indu Basti under Kumargram Block of Jalpaiguri District of West Bengal to identify demographic features, social culture and resource mapping and utilisation of Indigenous Fishing Technique by the Fisher women in the target villages. The main objectives of the present study are:

i. Assessment of available resources in the Rava Basti and Indu Basti area and identification of their age old gears and fishing techniques.

ii. Developing a future plan for better livelihood using sustainable aquaculture and captive breeding through formation of Cooperative Societies and SHG groups.

iii. Involving local people in the planning process through PRA.

MATERIALS AND METHODS

A study made during 2009-10 in a tribal village inhabited by Mech or Boro, 'Rava' and Santal tribe who collect wild specimens using various sorts of traditional gears. The 'Rava Basti' forest village at Barovisa G.P and Indu Basti Forest Village at Changmari G.P. is situated beside the bank of the river 'Jorai' near Silbangla forest in Jalpaiguri district of West Bengal.

Sampling survey was conducted in some pockets of the said area where native ornamental fishes are available. The pockets are fresh water zone and cold water zone of Jorai river.

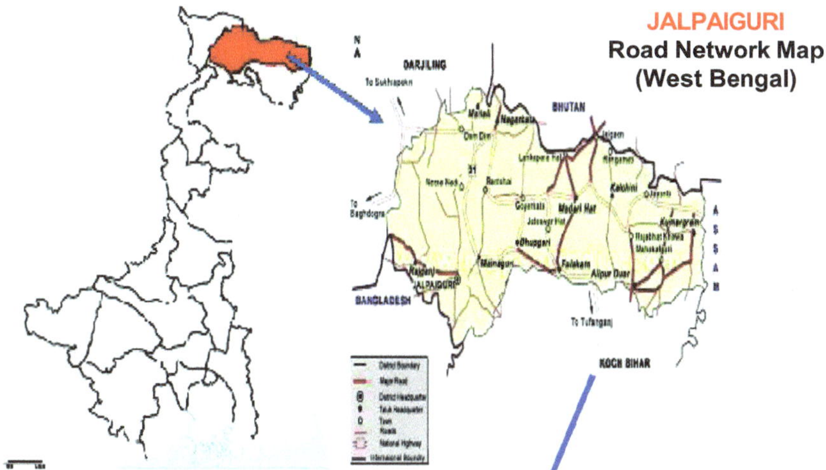

Fig. 1: Map showing survey area

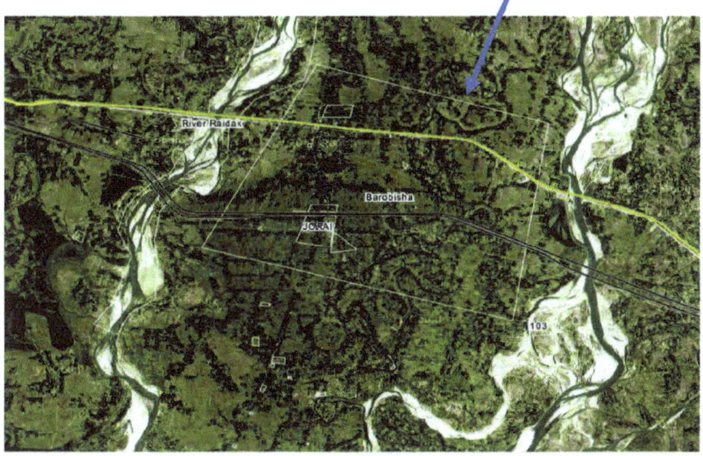

Fig.2: Satellite view of the study area in Kumargram block of Jalpaiguri district of West Bengal.

RESULTS AND DISCUSSION

Peoples participation is also recognized as an essential element in strategies for sustainable developmental in the process to reduce the cost to governments of providing development assistance. In 1970s dissatisfaction with the results of long-term studies prompted a search of more cost effective and above all more realistic methods/approaches with a thrust in i. Participatory Rural Appraisal, ii. Participatory Learning Method, iii. Participatory Action Research.

Role of the developmental workers is to be a catalyst, a facilitator and convener of the change processes within a community to help them to explore the present status and to come out with alternate ideas for improving their situation. According to Newton et al. (2006) participatory rural appraisal (PRA) evolved as a research approach from agricultural surveys. It involves local communities as active analysts of their own situation so the communities can set their own priorities on how to change their situations. This empowers people, giving them a sense of ownership in India. PRA techniques are now been used in different Government projects in different states like Andhrapradesh (Gona et al., 2009). Mascarenhas et al., 1991 reported the same type of practice in Karnataka, Tamilnadu and Andhra Pradesh. The survey of the Rava and Indu Basti forest village revealed many facts and figures.

Resource map

The resource map made by the participants is as follows. This is a complete picture of all the available resources in the villages. This will enable to comprehend existing resources as well as proper utilisation for sustainable development of an area.

Resource map shows clearly forest, plantation dwelling house and agricultural plots mostly located by the side of the canal connecting Jorai river (Fig. 3).

Fig. 3: Resource map

Fig. 4: Jorai River

Venn diagram

Venn diagram was constructed based on the resource survey carried out at the two villages through PRA technique. This is essential for any planning and developmental activities. The Venn diagram on the stake holders of the study made by the participants is shown in the diagram (Fig 5).

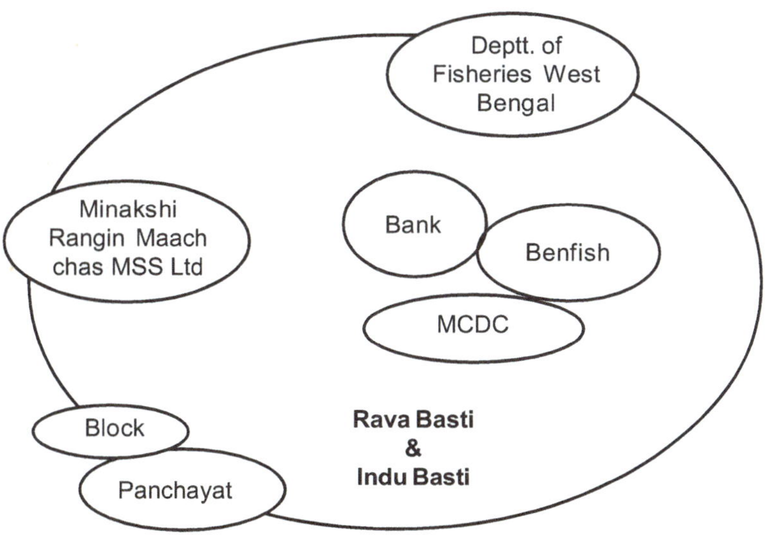

Fig. 5: Venn diagram

Resource utilization using traditional culture and value

Many communities are living in this area but most of them are the tribal people. The Ravas depend on the natural resources lying within their reach. The system of control and exploitation of land, water, and forest resources, the indigenous labour system, system of marriage by service, concept of totemism gradually going into oblivion, marriage celebrated at the boy's house, the belief that to the matrikin, the aspect of filation and such other cultural traits make the Ravas important from anthropological stand point. Befor their Hinduisation the tribal priest (Deosi) of the Ravas used to perform worship *viz., Nagu Kan Bakau* in the month of April, *Mahakal, Thansiri, Michil Bay tanga.* They performed *Matri Puja, Nur Tanga or Akash Kamini Puja* in case of child birth. However, these tribal people are facing various types of acute problems from different aspects. They are so poor and ignorant that they cannot receive their basic requirements of their survival. They have no alternative way to get rid of these problems. Though they are mainly depending on the agriculture for earning but this agricultural-job is a seasonal work. So they are facing disguised unemployment in the agriculture base work. There is no wide scope for earning from the seasonal agricultural based work. Naturally they will have to find other alternative source of income to maintain their livelihood. So the other main source of income of these people is to catch ornamental fishes from the adjacent rivers namely Jorai, Mansai and Raidak. They used to go for community fishing during season. At the time of community fishing (Bahamara) some four to five hundred people gather from the near by villages. In response to the sound of bugle of buffalo horn, they shout saying "*Buro thakur bole Hari Hari Bol* and after that only they start fishing in the river Rahidak which was previously known as Buranadi.

Women in ornamental fish collection, culture and trade

Mostly the female members of these traditional fishing communities of this region are engaged in ornamental fish capture. Children are also involved. Actually this is the primary occupation of these families. They generally go for fishing in group for all day.

Fig. 6: Tribal village of Rava

Fig. 7: Tribal women of Rava

Abundance of ornamental fish species

The study revealed that about thirty species of ornamental fishes are available throughout the year in the Jorai river. Among them only 10 to 15 species have market demand. At present the 'Chang' fish is very much available from these rivers and it has higher export value. But the catchers cannot get exchange-price properly. Earlier, the availability of the quantity of 'Sabuj chingri' and 'Rajabam/ Bamos' were much more but of late, its availability became lower than that of earlier. For example, earlier the availability of fishes per netting was @ 1000 pieces where as now a days it reaches @ 10 to 20 pieces. This may be due to spreading of poison like 'thiodon, copper sulfate' etc. by some dishonest fishers on the surface of the river water to kill the freshwater edible fishes for fulfilling their misdeed. As a result the quantities of the ornamental fishes are becoming less even if some species of ornamental fishes are on the verge of extinct.

Table 1: Various species of ornamental fish with local and scientific names are in the adjoining rivers

Local Name	Scientific Name
Raja baam	*Anguila bengalensis*
Katli	*Neolissocheilus hexagonalepis*
Gooch/Limbuni	*Bagarius bagarius*
Titakabri /Titay	*Balitewra brucei* (Gray)
Puti	*Barbus conchonius, B. ticto, B. stigma*
SoleTakiHile	*Channa striata, Channa punctata, Channa stewartii, Chana amphibious**
Jal Kapur	*Clupisoma montan**
Lohari /Utta	*Crossocheilus latius*
Chepti	*Cyprinion semiplotum**
Bhitti /Punti /Naptee	*Danio devario, Danio dangila, Danio aquipinnatus*
Til Kaprey	*Euchiloglanis hodgarti*
Kaprey	*Pseudocheneius sulcatus*
Buduna/Nak-katwa	*Garra gotyla, Garra annandalaiei, Garra lampta* (Hamilton)
Jhinge macha	*Macrobranchium* sp.
Teramas Tera	*Mastocembalus armatus*

*Critically engendered

Indigenous crafts and gears used for catching fish

Fisheries are well established activities in most part of West Bengal and the traditional knowledge and the expertise o the fishers have the potential to be utilized for bigger and more sustainable harvest. (Mukherjee et al., 2010). In view of the diversity in the nature of water bodies, fish fauna, socio-cultural ethos, indigenous technical knowledge and also availability of natural construction materials like bamboo, cane, etc. a large number of fishing gears are used in different areas. In the study area different gears are being used which are usually made up of split bamboo as given in the Figs. 8-12.

Fig. 8

Fig. 9

Fig. 10

Fig.11

Fig. 12

Fig. 8-12: Different local fish traps used in this region.

Marketing of ornamental fishes

Mainly the female members as well the children are engaged in collection of ornamental fishes and the male members are engaged to sale these collected fishes to the middlemen (broker) who deprive the catcher from their bonafied earning as because the fish catcher do not know exactly where and to whom their catch to be sent and sold. For this reason these innocent catchers will have to contact to the middlemen to sell their collecting fishes and in a single word, the fish catchers will have to surrender their collecting fishes to the middlemen with a very negligible price. As a result the original fish catchers earn a little money whereas the middlemen earn huge money. The fish catchers usually sell the ornamental fish @ 0.07 to 0.1 INR/piece whereas the middlemen sell the same fishes after purchasing from the catchers @ INR 0.60 to INR 1.50/piece to the exporter. It is noted here that the exporter's selling price is totally unreported because there is no concerned authority that regularly look after about the actual price by which the fishes would be purchasing or selling.

CONSTRAINTS

The present survey reveals that lot of constraints faced by the tribal fisher folk which are listed as follows:

- Excess sediment particularly during the rainy season increase turbidity of water resulting in decrease of food organism of the system.
- Soil erosion following deforestation results in siltation covering spawning sites with sediments that can ultimately become a cause of loss of fish diversity.
- Dams act as physical barriers for migration, preventing access to fishes to their usual breeding, rearing & feeding grounds.
- Indiscriminate use of nylon gill nets of small mesh size, attempt to catch all the fishes in the dry season, to catch fishes from breeding ground results in severe population disturbance.
- Killing of fish by pesticides (Chlorinated hydrocarbons, organophosphates and carbomates), copper sulphates, bleaching powder are rampant.
- The ornamental fish catchers have been depriving from their actual income due to the interference of the so-called middlemen.
- Non-availability of crafts and gears in time and space.
- No electricity availability, so establishment of hatchery could be rarely possible.
- No rhythmical availability of the quantity of fishes throughout the year.
- Lacking of proper training regarding culture, maintenance and transport of their produce.
- Communication gap between the exporter and the catcher.
- Non-availability of oxygen cylinder for maintaining their catch properly. They used to buy each pouch of Oxygen @ INR 5/-.
- Lack of budgetary support.
- Non availability of hatchery or seed bank.
- Poor infrastructure and undeveloped market linkage system.

CONCLUSION

From the above study it has been felt that application of PRA process is less time consuming and economic, but the information gathered from this process are not that accurate. The villagers though initially were little bit shaky but later most of them even the person who initially showed almost no interest was also got involved in the process. The resource picture gained through application of PRA when plotted in tables was very much informative about the Indigenous resource availability and very much helpful in formulating future plan village development. On the basis of the indigenous knowledge used by them, different developmental activities have been planned to improve the livelihood of the women fishers of this area. They are expert to use their traditional traps in more improvised way for better catch in good condition and get more return as live ornamental fish. To avail market facilities two cooperative societies have been established by the women with the help of Fisheries Department of West Bengal and National Cooperative Development Corporation (NCDC). To continue the trade round the year as well as to protect the natural biodiversity, beside wild catch they are being trained to make captive breeding of these small indigenous species aiming standardization and sustainable development of this area mostly inhabited by tribal dominated by women fisherfolk.

Fig. 13: Cement cisterns for rearing and captive breeding of small indigenous fishes

References & Recommended Readings

Gona, J., Xiong, K., Muhit, T., Mohammad, A. Newton, C. and Sally, H., 2009. "Identification of people with disabilities using participatory rural appraisal and key informants: A pragmatic approach with action potential promoting validity and low cost" Disability and Rehabilitation **32**(1), May 2010, pp. 79-85(7)

Mascarenhas, J., 1991. "Participatory Rural Appraisal and Participatory Learning methods: Recent experiences from Myrada and South "PRA Notes (1991), Issue 13 pp.26-32, IIED, London.

Mumherjee, M., Singh, S. and Mondal, S., 2010. Live feed Business for Ornamental Fish in Kolkata. Indigenous Technical Knowledge Practised in Fisheries Sector of Bengal Fisheries, West Bengal, pp-66, publication from Dept. of Fisheries, West Bengal.

Newton, C. R. J., 2006. Developing farm management type methods for participatory needs assessment. *The International Electronic Journal of Rural and Remote Health Research, Education, Practice and Policy.*

Chapter – 13

Selected Indigenous Technical Knowledge (ITK) on Fisheries in Tripura State

Sagar C. Mandal, Pampa Bhattacharjee and Debtanu Barman

INTRODUCTION

The origin of fish culture in India probably dates back to 2000 years. Possibly, fishery activity started when the human settlement moved away from the riverbanks to the hinterland. As the days went by, the Indian fish farmers practice their own indigenous technology as a common practice through natural resource management (National Commission on Agriculture, Gov. of India, 1976). The close and intimate observation of farmers through their age-old practices has developed the expertise of acquiring knowledge in aquatic ecosystem in particular. Admittedly, the multifarious knowledge and skill possessed by local people can substantially contribute to productive efforts and endeavors (Dasgupta, 1993).

The Northeastern region is one of the hot spot of biodiversity in the country and it is inhabited by about 30% of the tribal population, where more than 120 communities are distributed in 8 states (Islam et al., 2011). The North-Eastern region of India comprising of eight sister states namely, Arunachal Pradesh, Assam, Manipur, Meghalaya, Mizoram, Nagaland, Tripura and Sikkim is recognized by World Conservation Monitoring Centre (WCMC, 1998) as one of the global hot spots of freshwater biodiversity. Among these states, Tripura is one of the smallest states, which extends between 22° 56' N to 24° 32' N and

90° 09' E to 92° 10' E, covering an area of 10,492 sq. km. Tripura represents a transition zone between the Indian, Indo-Malayan and Indo-Chinese sub-regions of Oriental Zoogeographical region and is very rich in floral and faunal biodiversity (Gupta, 2000). The economy of this state is based on agriculture and allied activities in general and particularly, fisheries is considered to be one of the vital sectors for economic development. Fish farmers of the state have notable interest towards different issues of fish culture as experienced from last few decades. Farmers are innovative and skillful to use unique technical system in fish culture and its management. They are keen to observe the problems arising during farming and seek solutions to adjust to their local environmental conditions from generation to generation and generated an indigenous undocumented knowledge base. In Tripura, a small land lock state of North-Eastern region of India, the fishers have a rich legacy of indigenous knowledge, traditional practices and customary rules and regulations.

WHAT IS INDIGENOUS KNOWLEDGE (IK)?

According to Grenier (1998), the indigenous knowledge (IK) refers to the unique, traditional, local knowledge existing within and developed around the specific conditions of women and men indigenous to a particular geographic area.

Warren (1991a, b) has defined indigenous knowledge as the local knowledge that is unique to a given culture or society. Indigenous knowledge contrasts with the international knowledge system generated by universities, research institutions and private firms. It is the basis for local-level decision making in agriculture, health care, food preparation, education, natural-resource management and a host of other activities in rural communities.

According to Flavier et al. (1995), indigenous knowledge is the information base for a society, which facilitates communication and decision-making process. Indigenous information systems are dynamic and are continually influenced by internal creativity and experimentation as well as by contact with external systems.

Das et al. (2002) stated that indigenous knowledge refers to not only knowledge of the indigenous people but to that of any other defined community.

Wang (1988) defined indigenous technical knowledge (ITK) as the 'some of total knowledge and practices which are based on people's accumulative experiences in dealing with situations and problems in various aspects of life and such practices are special to a particular culture.

WHY INDIGENOUS KNOWLEDGE (IK) IS IMPORTANT?

Increasing attention indigenous knowledge is receiving by academia and the development institutions have not yet led to a unanimous perception of the concept of indigenous knowledge. In the emerging global knowledge economy a country's ability to build and mobilize knowledge capital, is equally essential for sustainable development as the availability of physical and financial capital (World Bank, 1997). The basic component of any country's knowledge system is its indigenous knowledge. It encompasses the skills, experiences and insights of people, applied to maintain or improve their livelihood.

The indigenous technical knowledge (ITK) is specifically concerned with the actual application of the thinking of the local people in various operations of agriculture and allied areas. The IK is based on experience, often tested over centuries of deeply embedded in local experience and historic reality and is therefore, unique to that specific culture and plays an important role in defining the identity of that community. Thus, it involves all the skills and innovations of people and embodies the collective wisdom and resources of that community.

It is acknowledged worldwide that traditional knowledge has advantages over other practices like indigenous knowledge has very minimum cost and is easily available to the farmers. The indigenous technical knowledge is socially acceptable, economically viable and sustainable and involves minimum risk to the rural farmers and producers and above all, they are believed to conserve resources (Grenier, 1998). The knowledge developed from experience, has enriched over centuries and has been embedded in local cultures and traditions handed down generations after generations by forefathers to the posterity for preservation and conservation of the repertoire of natural resources, soil, plant and other organisms (Lekshmi and Babu, 2009). So, there is an urgent need to understand, respect and tap these social resources for better governance of fisheries. Proper documentation of this knowledge will provide essential elements to formulate management plans towards ecologically sustainable development of fisheries. Even though, the scientific justification of all these technical knowledge is not available, but there is every possibility of having scientific base for theses ITKs. It is evident that the indigenous knowledge has lots of field applicability and it would help fisheries for its growth, development and future upliftment. So, there is urgent need for documentation of indigenous knowledge, their scientific discussion in the proper forum and validation, which is a big challenge in the present scenario.

SELECTED INDIGENOUS TECHNICAL KNOWLEDGE (ITKs) OF TRIPURA

1. Application of Banana Ash as a substitute of lime

The extract of banana ash known as "Kalakhar" locally, is rich in potassium. It is alkaline in nature and is often used in maintaining the pH level of the fish pond. The pH of water is an important criterion to produce adequate amount of natural food in the water through effective release of nutrients for the production of such organisms. Application of lime is essential to rectify pH of water. In rural areas, availability of lime may be a constraint. Hence, in place of lime, ash produced at home by burning banana leaves can also be used. Ash produced from banana is reported to be having very good impact in improving pH of water. As a general rule, in place of using one kg lime, three kg of ash would be required to have the required effect. Ash is also a good fertilizer and hence, farmers can be advised to use all the available ash for application in to fish pond.

Description of the ITK

Soil and water of Tripura are mostly acidic. Therefore, to increase the pH of pond water to an alkaline range, high dose lime is applicable. It makes good production of fish with enhanced growth. Banana plant ash contains high amount of potassium and it can be supplement to lime. Moreover, banana plants can be planted on the dike of the pond which will add extra source of income along with pisciculture. The banana ash can be produced by burning stem, roots and leaves of banana plant, prior to it banana plants are thinly sliced and sun dried slices are burned up to 80-90%, not completely.

The dose required for 1 kani (0.16 ha) is approximately 30-35 kg and the ash is spread along with the burned rice straw (*khar*). Farmers mostly used this practice during the winter seasons to get rid from EUS (Epizootic ulcerative syndrome) disease of fish.

2. Application of Durki plant/leaves to maintain good water quality

Green water is essential to increase the growth of fish based on the natural food produced in the pond. Most carps depend on the natural food produced in the pond and this is eaten by the fish through the filtration process. Northeast has plenty of greenery and vegetation that is rich in nitrogen such as Durki leaves. All such plants can be composted in various corners of ponds by creating bamboo fence and dumping the vegetation in such enclosures.

Description of the ITK

Durki plants are a common group of plant in Tripura, which grows rapidly in the village areas. This plant can be used as a source of manure along with cow dung to maintain good green water quality in the culture pond. The plant's leaves or stem can be cut and placed in the ponds corner with a bamboo fencing to stimulate the plankton growth in the system. Adding manure to vegetation pit would stimulate the process of composting in aquatic environment and increase productivity. Fish like silver carp, tilapia, that can filter plankton very efficiently grow very rapidly when such environment is created. Make sure that you do not over dump vegetation in the pond and which may cause oxygen depletion. Repeated applications are far more efficient than dumping at a time.

3. Application of Datura Fruits as a source of fish killing

Datura plants (*Datura wrightii*) are easily grown to the hilly and plain areas of Tripura. Datura plants gives oval shaped fruits which can be used as fish poisoning agents in the rural tribal areas of Tripura.

Description of the ITK

Datura plants' fruits are collected in large quantities and crashed. Crushed fruits along with the fruit juice mixed with water and spread all over the ponds. After 30 minutes to 1 hour, fishes start to come to the surface and there is a sudden depletion of dissolved oxygen in the water for which fishes start to die. It is easier for fishers to collect and apply this for fishing practices as Datura plants are locally available with the fish farmers and they does not require any additional expenditure to kill all the unwanted fish in the pond.

4. Control of Fish Lice (*Argulus*) using plam tree leaves and bamboo leaves substrate

Fish Lice (*Argulus*) is a most common ecto-parasite found in the fish culture ponds of Tripura. Fish lice caused Argulusosis disease which caused ulceration in the fish body along with the appetite loose which ultimately hampers the production of fish due to the hindered growth of fish. To get rid off it, farmers use palm tree and bamboo leaves as substrate in the fish pond.

Description of the ITK

Fish lice are very commonly occurring parasitic disease found throughout the Tripura state. Fish lice mainly occurred in the pond where there is a more stocking density or due to the high organic load presents in the pond ecosystem, which helps in the proliferation of the lice in the cultured ponds. When there is a attack of fish lice in the culture ponds fishes used to lose their body weight along with the ulceration in the body, which ultimately hamper the production and economic lose of the fish stock.

To prevent it to some extent farmers put palm tree leaves or bamboos leaves as substrate in the pond, where fish can move and rub their body (skin) against the hard/rough substratum to get rid off the lice.

5. Use of Refugee lata (*Mikania cordata*) as feed and manure to the culture fish pond

Refugee lata (*Mikania cordata*) has spread to moist tropical and subtropical regions and the North-East of India and known as invasion of alien plant species is native to Central and South America. This plant species can be found easily in the rural areas of Tripura. It has been observed that this plant species after chopped and mixed with cow dung as a compost can be used as a good source of plankton in the culture ponds and directly can be fed to the grass carp fish or directly can also be added to the rice bran to feed the fishes.

Description of the ITK

Refugee lata are most commonly used in the rural tribal areas as a source of feed to the fish and for manuring the fish pond. Refugee lata can be directly fed to the grass carp in the pond or can be composted with rice bran and fed to the fish or can be used as a source of fiber in the feed.

In other ways, Refugeee lata can be composted along with the cow dung and after decomposition can be spread to the pond which will accelerate plankton production or primary production in the fish pond, which will be beneficial for the filter feeding fishes and their growth.

6. Use of Plant Extract to control Epizootic Ulcerative Syndrome (EUS) disease

Epizootic Ulcerative Syndrome (EUS) is the most commonly occurring fish disease in the North-East region and even in Tripura, especially during winter months when the water level in fish pond is reduced considerably. The EUS is the most deadly ulcerative disease in fish caused great economic loss in the aquaculture system. The reddish coloured ulcer over the fish body is the external symptom that farmer used to identify it. Several traditional practices were recorded from farmers to control this disease. It occurs due to poor water quality management in the ponds. In winters due to low water temperature, EUS spread very easily. In rural tribal areas fishermen commonly practiced Refugee lata (*Mikania cordata*) and mixed it with green turmeric juice to control EUS disease in the fish pond.

Description of the ITK

The fishermen use Refugee lata and green turmeric juice together with ash for treatment of EUS in their ponds. They spread the ash (mixed with juice) throughout the pond. Ash is produced in the household by burning of fuels, banana leaves, straws and wood materials.

Green turmeric @ 15-20 kg for 1 ha pond is very effective. It is practiced mainly in the beginning of the winter and monsoon season. Some of the fishermen consider this method more effective, cost effective and eco-friendly. The scientific pre-requisite for control of EUS is to make the water alkaline. As the ash makes the water alkaline, it may be useful in preventing EUS. The spread of EUS shows that it involves the role of an infection agent in association with the water quality parameters. Thus, the application of leaf extract, ash and turmeric, which have antimicrobial action, have been reported to control EUS to a greater extent.

7. Usage of Milk Powder for Fish Seed Transportation

Good quality fish seed always play a vital role in the higher aquaculture production. Fish seed transportation from nursery to rearing tank is a crucial step in aquaculture practices. Farmers used traditional techniques to reduce the rate of mortality during transportation of seed. Most of the time, transportation of fish seed to distance places caused mass mortality due to stress to the fish seed during transportation. The main cause of mortality is due to the rise in water temperature in the hundi (*hundi* is aluminum container used by farmers' for fish seed transportation) because of direct sunlight. To overcome this problem, farmers used milk powder in the water to maintain low temperature inside the hundi.

Description of the ITK

This method mostly used by farmers in case of some non-hardy fish species like Mola *(Ambylopharyngodon mola)*, Chapila (*Gudusia chapra*) and many

Puntius species. During fish transportation from one place to another place, a small percentage of milk powder usually 2-3 spoon added to the water of 10 liter in the hundi to reduced the direct sunlight penetration to the water and make the water bit opaque and prevents direct heating.

8. Control of Tadpole and Snail by biological ways

Tadpole and snail always are destructive for aquaculture ponds. Tadpole creates problems in the tank by lowering the dissolved oxygen and even more responsible for poor water quality. Same ways, snails also create problem in fish harvesting and share in the food chain in the pond ecosystem, which ultimately affects in the overall aquaculture production. Even snails are also primary or secondary carrier of many fish diseases. To control these problems, farmers used some biological ways like growing of black carp and kept ripe jackfruit cover in the pond corner.

Description of the ITK

Tadpole and snails are growing in our aquaculture ponds during the monsoon time. Farmers used to stock @7-10 numbers of Black carp for 1 ha pond to control snail. Black carp is a snail eating fish.

To control tadpole they used to keep ripe jackfruit covered (after opening the whole jackfruit) near the pond margins or sides where this tadpole used to aggregate and stick to this substance. Then the farmers can remove the tadpole from it very easily.

9. Usage of Milk Powder as an Anti-Sticky Substance

Many catfish eggs are very sticky and adhesive in nature such as Pabda (*Ompok bimaculatus),* Singhi (*Heteropneustes fossilis*), Magur (*Clarius batrachus),* etc. Due to this adhesiveness, most of the time catfish eggs are susceptible to fungal diseases. Farmers used a small percentage of milk powder to remove the eggs stickiness.

Description of the ITK

Catfish breeding always needs more careful handling compared to carp breeding. Catfish eggs are sticky in nature for which they stick to each other like a bunch of grapes. To mix the eggs properly with milt, feathers are usually used but due to the eggs stickiness it always creates trouble in the mixing. Adding half or full spoon of milk powder in the eggs solution can resolve this problem and can be mixed the eggs properly in regards to fertilization.

10. Juck Fishing (Fish Aggregating Device)

Juck fishing is a special technique used by the fishermen in Rudrasagar Lake of Tripura to catch fish. In this type of fishing, 12 to 25 m^2 water area is surrounded by bamboo and its branches in the water body to make either circular or square shape and inside that circle, different aquatic weeds are put mainly *Eichhornia*, which become very suitable place for fishes, which provides shelter as well as provides food for the fishes. During harvesting whole juck is encircled by net and normally it takes 2 to 3 days to harvest fish. The most commonly catch in the juck are common carp and rohu along with many small group of fishes likes puntius, tengra, pabda, etc which is a common old practice in the lake mainly in winter seasons.

Description of the ITK

In the Juck fishing, deeper areas of the lake are chosen and the respective place are prepared by putting tree or bamboo branches on it and surrounded by water hyacinth (*Eichhornia* sp.). The place guarded by bamboo poles so that aquatic weeds and juck materials are not drifted away. The fish start to accumulate this kind of shelter in search of food and get protection from carnivorous fish or other predatory animals. Fishermen used to provide artificial food in the juck like household food waste material mixed with dry fish powder and rice bran. Feeding is not done every day; it's used to follow once in a month or twice in a month. After 2-3 months the juck area is encircled with the net and gradually fishermen moves forward to shorten the areas in view to catch the fish efficiently using bamboo poles. Meanwhile, the water hyacinth and other tree branches are removed from the harvesting area. The net is lifted slowly from all sides and fish are caught by hands or by scooping. Juck fishing is a kind of community fishing where 8-10 numbers of fishermen are required in the whole operation.

DISCUSSION

The Indigenous Technical Knowledge (ITK) has been widely used in the state of Assam on fish pond construction, pond management, fish seed transportation and fish health management were perceived high on cost effectiveness and easy to follow, moderate on scientific value/logic, efficacy, availability of material, cultural appropriateness and environmental soundness (Kalita, et al., 2004). These authors also reported that farmers of Assam hills perceived ITK because of their cost effectiveness and easy to follow. Few ITKs related to fish health management and fish seed were common both in Assam and Tripura state. Therefore, it is imperative to analyze these technologies so that the scientific principles/basis behind them could be properly understood. Once this is done, it will be easier for extension of ITK, further refinement and upgrade them by blending them with the modern scientific knowledge.

Proverbs are yet another means of communicating ancient wisdom from generation to generation. Proverbs of the fishing communities are no exception to this and have great significance and implications for day-to-day life. Such proverbs and traditional wisdom should be preserved and conserved to enrich the learning experiences of the present and future generations (Lekhsmi and Babu, 2009). The available ITK are based on the trials performed by their ancestors since generations and found effective in rural settings where timely extension services are not available in emergency cases. Investigation of scientific logic behind every ITK is imperative to study for better utilization and management of ITKs. Such scientific investigation would be quite useful to properly understand the logic behind every ITK and it provides the scope for refinement in the technical knowledge (TK) so as to upgrade it scientifically and extent it for betterment of the farmers community (Das, et al., 2013). Experimental trials on ITK would provide the scope to identify the cost-effectiveness, accessibility, compatibility and sustainability. Finally, amalgamation of such ITKs with scientific knowledge may upgrade it for enormous benefit for society through economic development on a sustainable way.

At present, many of the indigenous technical knowledge's (ITKs) are at risk of becoming extinct because of rapidly changing natural environments and fast economic progress, political and cultural changes on a global basis. As ITKs become inappropriate for new challenges or because they adapt too slowly, they vanish from the society. Many practices of ITKs disappear only because of the intrusion of foreign technologies or development concepts that promise short-term gains or solutions to problems without being capable of sustaining them for future. The tragedy of the impending disappearance of indigenous knowledge is most obvious to those who have developed it and make a living through it. But the implication for others can be detrimental as well, when skills, technologies, problem solving strategies and expertise are lost forever.

CONCLUSION

The indigenous technical knowledge and learning about it is a common sense. It is a must to know knowledge for all to maintain a balance in their life physically, mentally, spiritually and from their heart for the goods of humanity and the betterment of their community and the society at large. All the above mentioned indigenous technical knowledge was found to be viable in combating the problems of the aquaculture and related fields in North-Eastern region as a whole and Tripura state in particular. Fish farmers knowingly or unknowingly employed such techniques in effective management of them. The scientific approach of these techniques can direct a new era in modern fish farming system employing these indigenous technical knowledge for better aquaculture production on sustainable basis.

References & Recommended Readings

Das, A., Debnath, B., Choudhury, T. G., Roy, D., Saha, B. and Paul, A., 2013. Indigenous technical knowledge for pond maintenance, fish health management and fish seed in Tripura, India. *Indian Journal of Traditional Knowledge* **12**(1): 66-71.

Das, P., Das, S. K., Arya, H. P. S., Reddy, G. S. and Mishra, A., 2002. Inventory of Technical Knowledge in Agriculture. Mission Mode Project on Collection, Documentation and Validation of Indigenous Technical Knowledge Document-I, ICAR, p. 3-8.

Dasgupta, D. D., 1993. Traditional wisdom and experience in agro technology generation and use, XVII Indian Science Congress, Bangalore.

Flavier, J. M., Jesus, A.de, Navarro, C. S. and Warren, D. M., 1995. The regional program for the promotion of indigenous knowledge in Asia", pp. 479-487, In: Warren, D. M., L. J. Slikkerveer and D. Brokensha (eds.) *The cultural dimension of development: Indigenous knowledge systems*. London: Intermediate Technology Publications.

Grenier, L., 1998. Working with indigenous knowledge: A guide for researchers. Ottawa: IDRC. In. *Conservation of Biodiversity: Through Indigenous Knowledge*, (Eds.) Paul S. and Ramanathan, A. 2002, *Kurikshetra* **5**(3): 3-7.

Gupta, A. K., 2000. Shifting Cultivation and Conservation of Biological Diversity in Tripura, Northeast India. *Human Ecology* **28**(4): 606-6629.

Islam, S., Datta, M., Sankaran, M., Prakash, J., Das, S. P., Biswas, S., Sulabidya, A. and Bhattacharjee, T., 2011. Underutilized Pomological Crops of Tripura- An unexplored potential, *Tripura Science Congress*, 2011, pp.22.

Lekshmi, P. S. S. and Babu, A. P. D., 2009. Indigenous Technical Knowledge and ancient proverbs of the coastal fisher folk of Kerala and their implications. *Indian Journal of Traditional Knowledge* **8**(2): 296-297.

Kalita, B., Choudhury, M. and Ojha, S. N., 2004. Indigenous technical knowledge on pond construction and maintenance, fish seed transportation, and fish health management in Assam hills. *Indian Journal of Traditional Knowledge* **3**(2): 192-197.

National Commission on Agriculture, Government of India, Ministry of Agriculture and irrigation, New Delhi, part VIII (Fisheries), 1976.

Wang, G., 1988. Indigenous communication systems in research and development, *J Ext Sys*. **4**: 75-86.

Warren, D. M., 1991a. Using Indigenous Knowledge in Agricultural Development; World Bank Discussion Paper No.127. Washington, D.C., the World Bank.

Warren, D. M., 1991b. The Role of Indigenous Knowledge in Facilitating the Agricultural Extension Process. Paper presented at International Workshop on Agricultural Knowledge Systems and the Role of Extension. Bad Boll, Germany, May 21-24, 1991.

WCMC, 1998. Freshwater Biodiversity: A preliminary Global Assessment Document. A document prepared for the 4th Meeting of the Conference of the Practices to the Convention of Biological Diversity, World Conservation Monitoring Centre.

World Bank, 1997. Knowledge and Skills for the Information Age. The First Meeting of the Mediterranean Development Forum; Mediterranean Development Forum, URL:http://www.worldbank.org/html/fpd/technet/mdf/objectiv.htm.

Chapter – 14

Indigenous Disease Management Practices in Traditional Farming Systems in India

A. Chakraborty and Subham Dey

INTRODUCTION

The enemies of plants have been harming the organized agriculture ever since man started depending on plants for food. Fossil evidence indicates that plants were affected by disease since 250 million years ago. Conditions of food shortage and famines have been created by these enemies of plants. Naturally, the man had been obvious of their presence and proposed explanations and methods of combating them according to the knowledge available at that time all over the world. It is not true that attention to plant diseases was first given in the western countries. In the ancient Indian literature many plant diseases and their remedies were mentioned.

Traditional farming system is an ecologically based age-old farming system using self reliance locally available resources without external inputs developed by ancient farmers through generations of their interaction for food, fodder and fibre. Indigenous farming system, once prevalent all over the world is now almost vanished from the developed countries and confined to some tribal dominated regions of developing countries occupying more than half of arable land (Thurston, 1992).

India is rich in traditional farming systems because of diversity in agro-ecological habitats inhabited by diverse ethnic groups who have been practicing age-old farming in location specific situations since generations. It is a community based farming system that brings the local people closer and respects their environment (Chhetry and Belbahri, 2009).

INDIGENOUS METHODS STARTING FROM ANCIENT ERA

Probably, development of agriculture started in the orient. There is mention of Chinese practicing crop rotation as early as 3000 BC and in the first century BC they were supposed to keep the field fallow for a year if the field gave poor yield in the second year. Seed health was mentioned in China by Fan Sheng-Chih in the first century BC. In the ancient western literature (the Greek and Roman civilization) the central figures of observations on plant life and plant diseases were Theophrastus (c. 372-287 BC), a Greek philosopher and writer on various subjects. Although most of the work of Theophrastus is lost, two of his books, *Historia Plantarum* and *De Causis Plantarum,* still find a place as reference. In his writings, Theophrastus elaborately mentioned plant diseases (rusts, mildews, blight, etc.) but expressed the opinion that these diseases were due to bad nutrition and bad air. There was no mention of association of living organisms with plant diseases. Others during that time (Varo, 116-127 BC, Maro, 70-19 BC) did mention specific treatments of seed.

In the Indian subcontinent, more than 3 centuries before the Christian era, during the Mauryan Empire (c. 320 BC), agriculture was fairly developed and was the main concern of the government. Kautilya, also known as Chanakya and Vishnugupt, had written (c. 320-296 BC) the classic treatise Arthashastra (Science of Source of Livelihood) which incorporated many observations and recommendations for healthy crop culture.

Much earlier (about 4700 BC and later) than the time of Theophrastus in the west and Kautilya in India sages had composed the four Vedas (religious verses). These were composed and, in absence of a script, were passed on through word of mouth from generation to generation for centuries. Rigveda (4700 BC) is considered the oldest composition of religious hymns. It was followed by Yajurveda (hymns and rituals), Samveda (Rigveda recomposed for singing) and Atharvaveda (3000 BC and later) which contained charms and spells for warding off evils and diseases. This composition specifically mentioned blight as a disease and its control. Ploughs and other agricultural implements were in use. In addition to such diseases as blight, powdery mildew, rust and tumours on trees, fungi (mushrooms) and algae were also mentioned in the Vedas. In Rigveda a number of verses were devoted as prayer to the Sun God for purifying and protecting everything and for destroying the tiny, invisible or visible creatures that poisoned the food and caused disease. Similarly, fire *(Agun)* was worshipped for destroying the poisonous beings and purifying articles used in religious activities. Obviously, thousands years before the time of Theophrastus the civilization in India was aware of living beings *(Krimi)* that caused disease in man, cattle and plants and also spoiled in the Vedas. These were visible and invisible creatures entered the body and caused disease and sun heat and fire killed these poisonous creatures.

More specific references to plant health and disease were found in written literature after the four Vedas when attempts were made to prevent and cure diseases. Susruta, the great Indian pioneer in medicine and surgery who wrote Susruta-Samhita (c. 400 BC) and Charak, also a man of medicine, were aware of diseased conditions in plants and often compared them to diseases in man. Although, Susruta did not deal with plant disease, while advising men for good health, he wrote "just as the proper season, good soil, water and vigorous seed produce a healthy plant". The oldest text on Indian agriculture- Krishi Parashar- was probably written by Parashar (c. 400 BC), before the Arthashastra of Kautilya. His book was considered the first extensive coverage of ancient agriculture of the Aryans. It contained chapters on most aspects of crop culture from sowing to harvest and seed storage. However, plant protection was mentioned directly in one verse that mentioned powdery mildew, rust, insects, and larger animals as enemies of crops and invoked the Wind God to move them away from his field. Parashar had declared that origin of plentiful yield was the seed, implying seed health. Kautilya (321-296 BC) in his Arthashastra, listed treatments of seed in addition to recommendations for punishment against sale of spurious seed.

More extensive coverage of ailments of plants was found in the text Vriksharyurveda by Surapal (c. 1000 AD). This text mainly dealt with cultivation of fruit and flower trees. It was based on previous compositions and writings as well as author's own observations. It covered such topics as importance of trees, their location, soil types, methods of propagation, tree nutrition, diseases and their treatment (Sadhale, 1999). Plant protection was already recognized as an important agricultural activity when Surapal wrote his Vrikshayurveda. Surapal was also a physician and had divided plant diseases, like human diseases, into two categories, internal and external. The internal diseases were caused by inroad of foreign organisms in the plant body while the external diseases were attributed to non-parasitic injuries by heat, frost, high winds, soil acidity, water stress, poor quality seed, etc. To avoid both internal and external diseases Surapal prescribed treatment of pits for planting trees, treatment of seed and treatment of standing trees (Sadhale, 1996).

INDIGENOUS TRADITIONAL KNOWLEDGE (ITK) IN PLANT DISEASE MANAGEMENT IN INDIA PARTICULARLY IN NORTH EASTERN INDIA

Indigenous knowledge is the knowledge of the indigenous people inhabiting different geographical regions of the world with their own language, culture, tradition, belief, folklore, rites and rituals. In course of their close interactions with nature and natural resources, they were to make a certain decisions as to the solutions of their problems they encountered in their day to day life while managing the land and environmental resources for survival. Compelling situations motivated them to generate knowledge out of necessities. Therefore, indigenous knowledge so developed was based on necessities, extinct, curiosity and observations of ethnic groups to mitigate the immediate situations (Chhetry and Belbahri, 2009).

This knowledge has been tested using the thumb rule of trial and error methods over a period of time through generations and validated to make the established knowledge for the purpose for which it was designed for use of the ethnic group who propounded for the first time. Thus, indigenous knowledge varies from tribe to tribe and usually passes on to the next generations though the words of mouth, actions or even practices, usually by the elders of the family/society. Eventually, this local knowledge in course of time got socially accepted and validated which finally entered into the social life and subsequently became the Indigenous traditional knowledge (ITK) of the society as a whole. Thus, socially validated indigenous knowledge which shared common values and gains popularity with human element attached to it particularly the culture of the ethnic groups might be known by the name Indigenous traditional knowledge. However, now-a-days it was known to people synonymously by various modern names such as indigenous technical knowledge, peoples knowledge etc. (Singh and Sureja, 2008).

With the advent of modern agriculture, traditional farming in India was largely confined to hilly regions that too in peripheral states where green revolution could not influence the traditional farmers. Traditional farming system in north east of India was complex and unique of its own as the land use system in this regions was dominated by slash and burn agriculture where management of crops and pests were carried out by integrating indigenous knowledge and traditional ecological knowledge of the communities (Chhetry and Belbahri, 2009).

North eastern region is one of the most picturesque regions having varieties of ethnic cultures. It is also diverse in ecological habitats ranging from mountains to valley with rich diversity in plants and crops and ethnic diversity. Ethnic communities of the region practice their traditional cultivation practices of diverse crop suitable in different ecosystems solely on organic conditions without any external inputs to the system; and value the land and the environment for the sustainability of the system for future generations (Chanu et al., 2010). They use indigenous traditional knowledge for controlling the menace of diseases and pests of the crops some of these traditional methods includes uprooting the infected plants, altering the crops and plants architecture, biological control, adjusting crop density, time of planting, fallowing, mulching, multiple cropping, planting without tillage, planting in raising beds, rotation. These traditional practices developed a balance for meeting their present needs and protecting the environment for future generations.

VARIOUS ITK PRACTICES AND THEIR RELEVANCE IN MODERN AGRICULTURE

Seed treatment

Interestingly, Kautilya had recommended that seed should be left in open for 7 or more nights and days. This, probably, was the oldest reference of seed treatment with solar radiation after wetting by dew in the night. Some other recommendations of Kautilya that could be viewed in the light of modern science included (i) sugarcane seed cutting end should be treated with lard and honey and pasted with cow dung and (ii) cucurbits should be cultivated on river beds during summer, a practice that had sustained for more than 2000 years and is still followed. Varahamihir (505-587 AD) in his Brihat-Samhita included a chapter on science of plant life. Apart from writing about fungi (mushrooms) and algae, he also advocated the importance of good seed and seed treatment for good and healthy seedlings.

The materials used for treatments of seed, field crops and trees during the period of Parashar to Surapal appeared to be exclusively organic sources. The list included cow dung, clarified butter (ghee), decoction of root of five trees, mustard, hog fat, cattle horn, milk, honey, liquid manure, oil of Madhuca, ash, cow urine, human urine, beef, extract of *Ernbeiia ribes* (most commonly recommended) and so many other things. These were available in plenty those days and farmers could use them at no cost. They may looked strange but modem science is gradually explaining their utility and this text. The application of cow dung to seed, pits, and tree trunks is even now practiced in certain areas and is effective.

Crop rotation

In the south, the Tamil poet Tholkappier (200 BC) considered plants as living beings, mentioned monocot and dicot plants, and wrote about benefits of rice-legume rotation.

The use of rotation should be carefully investigated and utilized in schemes designed to aid traditional farmers, keeping in mind that the value of crop rotation for the management of specific diseases is highly location specific. In addition to agronomic benefits, such as improved soil texture and better use of nutrients and water, rotations contribute to the management of soil borne pathogens. Rotations are probably the best documented traditional disease management practice.

Mulching

The list of materials used as mulches by traditional farmers is very long (Rowe-Dutton, 1957, Wilken, 1987, Wilson and Akapa, 1983). Cereal straw and stalks (Fig. 1) were perhaps the most commonly used mulches, but other examples were crop debris, sawdust, leaves, grass, corn stover, manure, weeds, reeds, Spanish moss, and various aquatic plants.

Fig. 1: Mulching with Cereal straw and stalk

Wilson and Akapa (1983) wrote: "Mulches also decrease soil moisture evaporation, increase infiltration rate, smother weeds, lower soil temperature, and enrich soils." Mulches reduce rain splashing, an important means of dissemination for numerous bacterial and fungal pathogens. Soil temperatures are lower under mulches in warm tropical areas. Such temperature changes can have significant effects on the ability of soil borne plant pathogens to cause disease.

Many authors suggested that mulches may reduce the incidence and severity of plant diseases. Rowe-Dutton (1957) made a comprehensive review of the literature on disease management using mulches for a variety of vegetable crops. Much of the information she cited was anecdotal rather than experimental. Mulches contributed to disease management in various ways. Reduction or prevention of soil splashing was an important function of mulches in the management of some plant pathogens (Fitt and McCartney, 1986, Gilbert, 1956, Galindo et al., 1983b, Moreno and Mora, 1984, Rowe-Dutton, 1957). According to Fitt and McCartney (1986): "Rain splash is the second most important natural agent, after wind, in the dispersal of spores of plant pathogenic fungi". Mulches may prevent direct contact of the foliage, fruit, or vines with the soil and thus prevent diseases transmitted from the soil. The severity of the disease blossom end rot of tomatoes could be reduced by mulching (Rowe-Dutton, 1957).

Organic mulches might have similar effects. Some information is found regarding the use of mulches as a covering layer by traditional farmers, but much more information is found on the use of various organic amendments incorporated into the soil. When organic mulches are so incorporated, there is considerable evidence that they usually have a suppressive effect on plant pathogens and nematodes (Cook and Baker, 1983). Organic amendments have been found to be useful in managing nematodes (Castillo 1985, Muller and Gooch, 1982). According to Hoitink and Fahy (1986) composts prepared with tree barks and used as mulch may release inhibitors of plant pathogens such as *Phytophthora* spp. and some nematodes.

Gindrat (1979) listed many examples of soil borne pathogens managed by the addition of organic matter, but noted that in many cases huge amounts of organic matter were needed to provide control. There are many soil borne plant pathogens, and the management of each may depend on different site-specific environmental and soil factors.

Mulches reduce plant diseases by reducing soil splashing, influencing the moisture content and temperature of the soil, and enhancing the microbiological activities that suppress plant pathogens. In the hot humid tropics, where plant growth is rapid and luxurious, the use of green manures and natural vegetation as mulches (as in the slash/mulch system) should be considered, as their use provides an effective management practice for some diseases.

Burning

Fire and heat are often overlooked as plant disease management practices. The high temperatures produced by burning may eliminate inoculum of many pathogens in the slash and burn farming system. The negative effects of burning should not be overlooked when contemplating the use of burning for plant disease management.

Organic farming and bio-rational control

On account of the inherent hazardous effect involved in conventional chemical management, the alternate plant protection measure like use of organic farming and animal products such as cow urine, cow milk and buttermilk as described in Vedas, Arthashastra, Agnipuran, Surapala's Vrukshayurveda (Sadhale, 1996) etc., are gaining importance. Cow urine and buttermilk are being used for the management of insect pest and diseases in many crops, which are named as antifungal, organic gold, biodynamic, bio-pesticide, bio-organic manure, bio-growth promoter and indigenous to know the mechanism of disease control in leaves of soybean as influenced by rust.

The recommendation of Vrukshayurveda suggests value of green manuring and organic amendment which are now known to reduce root diseases. Sesame is also a trap crop for *Striga* (Sadhale, 1996).

The value of green manure crops in agriculture has been known for centuries. Cato, a Roman who lived in 234-149 B.C., wrote that lupines, beans, and vetch fertilized the land. Varro, writing between 116-127 B.C., suggested that some plants, although they gave no benefit the year they were ploughed under, gave benefits the following year. Green manure has been used for centuries in China, and its present and past use has been mentioned by Cook and Baker (1983).

The practice of using green manure crops has important benefits, as they add organic matter to the soil, and play a role in the suppression of soil borne pathogens while improving the physical condition of the soil. Nutrients may also be added to the soil, especially from leguminous green manures. A wide variety of plants have been used as green manure crops (Karunairajan, 1982). *Crotalaria spectablis* is often used as a cover crop and subsequently ploughed under as a green manure. Root-knot nematodes (*Meloidogyne* spp.) enter the roots of Crotalaria, but do not survive. Thus, Crotalaria also acts as a trap crop and can be useful in the management of nematodes.

Flooding

Traditional agriculture has utilized flooding (Fig. 2) extensively for the management of plant pathogens. For example, the paddy rice system, in addition to its various agronomic benefits, has an important role in reducing the importance of soil borne diseases. Where feasible, flooding should be recommended as a possible plant disease management practice.

Fig. 2: Flooding, a common indigenous disease management practices

Multiple cropping

It is difficult to generalize with any degree of accuracy about disease management through the use of multiple cropping. Recommendations on multiple cropping should be thoroughly tested, and site-specific recommendations will often be necessary. Time tested local practices should serve as the first guide to recommendations. Nevertheless, most of the literature indicates that there is less disease in most types of crop associations than in monoculture.

Using organic amendments

Historically, many sustainable agricultural systems incorporated large quantities of organic matter into soil. This incorporation generally resulted in less soil borne disease, in addition to other important agronomic benefits, and the practice should be recommended whenever feasible. An interesting recommendation of Varahamihir was that when a piece of land is brought under a crop, sesame should be planted, chopped down before seeds are formed and incorporated in to the soil.

In the Kumaon Hills (Uttarakhand) the apple tree owners apply a mixture of cow dung and mud to cut ends of branches after annual pruning which protects the wound from infection until self healing. The extracts of decomposed excreta of animals have been found to reduce incidence of powdery and downy mildews of grapevine in Germany and other countries. Plant oils have been used effectively against powdery mildew of apple giving as much as 90% control.

Skimmed milk and cow urine have been demonstrated as disease control materials during the last few decades. Mechanically emulsified rape oil applied as spray is comparable to the use of Karathane against apple powdery mildew (Northover and Schneider, 1993) and grapevine powdery mildew (Azam et al., 1998). Applying a coat of pure mustard oil to ripe mango fruits gives 90% control of Aspergillus rot. Skimmed milk has been found a good seed treatment material against common bunt of wheat.

Fallow

Fallow periods are often beneficial in reducing losses from plant diseases, especially soil borne diseases. For disease management, fallowing is generally more effective in combination with rotations. Both dry and flood fallowing should be considered in planning plant disease management.

Fallowing has been practiced for thousands of years. In what is now Iraq, the ancient Sumerians, one of the world's oldest civilizations, practiced fallowing for cereal fields. A sabbatical year was mandated in ancient Jewish religious law; in that for one year farmers were not allowed to plant crops. Exodus 23:10-11 reads: "For six years you shall sow your land and gather in its yield; but the seventh year you shall let it rest and lie fallow".

Planting in raised beds

Raised fields, raised beds, ridges, and mounds were used widely by traditional farmers for millennia. Better drainage and irrigation, enhanced fertility, and frost control are other important benefits of these systems, but planting in soil raised above the soil surface is also an important disease management practice for soil borne pathogens.

Crop diversity

Many of the major crops on which humans depend for food are constituted primarily of cultivars or landraces selected before "modern" agricultural science began. These landraces are usually genetically diverse and in balance with the environment and endemic pathogens. They are dependable and stable in that, although not necessarily high-yielding, they yield something under all but the worst conditions. The conservation and possible utilization of these landraces should be considered a priority in disease management schemes.

Storage practices

The many ingenious and effective practices for crop harvesting and storage developed by traditional farmers over the centuries warrant careful study and evaluation, as some of their practices might be more practical and appropriate for traditional conditions than the storage practices current in modern agriculture.

Farmers found that by using this practice the grain is protected from rain, it dries better on the plant in the sun than in storage, is less likely to be blown down by the wind, is less accessible to rats and birds, and reaches a low moisture content that greatly reduces storage deterioration.

Limitations

The science of plant pathology has an important role in the future success of programs and policies designed to increase and sustain food production. If plant pathologists are to be effective in addressing the problems of food production in developing countries, the agricultural systems of traditional farmers in those countries must be thoroughly understood. Then, researchers can appropriately address problems in the context of the farmers' systems and efficient, proven techniques can be disseminated to other farmers. Traditional knowledge can be overvalued or romanticized, but it is a mistake to despise or ignore it. Far too many giant development projects in developing countries have failed dismally, often with serious ecological consequences, because sufficient understanding of traditional agriculture was lacking. Today there are many concerns about modern agriculture because it is highly energy-intensive and has a narrow genetic base and because attainment of increasingly higher yields and greater efficiency may lead to monoculture and overproduction. Sometimes destructive erosion, pollution, and excessive pesticide residues result. It is time to reexamine the potential for traditional agriculture to contribute to an improved, sustainable system for production agriculture.

Lack of proper scientific knowledge about causal pathogen and mechanism of management practices sometimes indigenous disease management practices fails. Some organic decomposition products are phytotoxic (Linderman, 1970, Patrick et al., 1964), and thus amendments are not always beneficial. Unfortunately, mulches provide a good environment for the multiplication and survival of slugs, which sometimes cause serious losses to crops such as beans when mulched. In Costa Rica the same slugs that attack beans also vector a serious human nematode pathogen (Beaver et al., 1984). Mulches may also provide nutrition and a suitable environment for certain plant pathogens.

References & Recommended Readings

Azam, M.G. N., Gurr, G. M. and Magarey, P. A., 1998. Efficacy of a compound based on canola oil as a fungicide for control of grapevine powdery mildew caused by *Uncinula necator*. *Australasian Plant Pathology* **27**: 116-118.

Beaver, P. C., Brenes, R. and Vargas Solano, G., 1984. Zoonotic filaria in a subcutaneous artery of a child in Costa Rica. *American Journal of Tropical Medicine and Hygiene* **33**(4):583-585.

Castillo, M., 1985. Some studies on the use of organic soil amendments for nematode control. *Philippine Agriculturist* **68**:1-18.

Chanu L. B., Chhetry, G. K. N. and Sharma, G. D., 2010. Sustainable indigenous practices for the management of pest and diseases of upland rice in Manipur, North east India. *Assam University Journal of Science & Technology: Biological and Environmental Sciences* **5**(1): 58-62.

Chhetry, G. K. N. and Belbahri, L., 2009. Indigenous pest and disease management practices in traditional farming systems in north east India - A review. *Journal of Plant Breeding and Crop Science* **1**(3): 28-38.

Cook, R. J. and Baker, K. F., 1983. The Nature and Practice of Biological Control of Plant Pathogens. American Phytopathological Society, St. Paul, MN. pp. 539.

Fitt, B. D. L. and McCartney, H.A., 1986. Spore dispersal in splash droplets. *In*: Water, Fungi and Plants. (Eds. P.G. Ayres and L. Boddy). Cambridge University Press, Cambridge, UK.

Galindo, J. J., Abawi, G. S., Thurston, H. D., Galvez, G., 1983. Effect of mulching on web blight of beans in Costa Rica. *Phytopathology* **73**: 610–615.

Gilbert, J. C., 1956. Soil mulches of local materials. *Hawaii Farm Science* **4**: 4-5.

Gindrat, D. (1979). Biological soil disinfestation. *In*: Soil disinfestation. (Eds. D. Mulder). Elsevier, Amsterdam. pp. 251-287.

Hoitink, H. A. J. and Fahy, P. C., 1986. Basis for the control of soil borne plant pathogens with composts. *Annual Review of Phytopathology* **24**:93-114.

Karunairajan, R., 1982. Green manuring in the tropics. *In*: Basic technics in ecological farming (Eds. S. Hill). Birkhuser Verlag, Basel, Switzerland.

Linderman, R. G, 1970. Plant residue decomposition products and their effect on host roots and fungi pathogenic to roots. *Phytopathology* **60**:19-20.

Moreno, R., and Mora, L., 1984. Effect of cropping pattern and soil management on plant diseases. II. Bean rust epidemiology. *Turrialba* **34**: 41-45.

Muller, R. and Gooch, P. S., 1982. Organic amendments in nematode control: An examination of the literature. *Nematropica* **12**: 319-326.

Northover, J. and Schneider, K. E., 1993. Activity of plant oils on disease caused by *Podosphaera leucotricha*, *Venturia inaequalis*, and *Albugo occidentalis*. *Plant Disease* **77**:152-157.

Patrick, Z.A., Toussoun, T.A. and Koch, L.W., 1964. Effect of crop-residue decomposition products on plant roots. *Annual Review of Phtopathology* **2**:267-292.

Rowe-Dutton, P., 1957. The Mulching of Vegetables. Commonwealth Agric. Bur. Tech. Comm., no. 24, Farnham Royal, England.

Sadhale, N., 1996. Surapal'sVrikshaayurveda (The Science of Plant Life by Surapai). Agri-History Bull. No.1. Asian Agri-History Foundation, Secunderabad, India.

Sadhale, N., 1999. Krishi-Parashar (Agriculture by Parashar). Agri-History Bulletin No. 2. Asian-Agri-History Foundation, Secunderabad, India.

Singh, R. K. and Sureja, A. K., 2008. Indigenous knowledge and sustainable agricultural resources management under rainfed agro-ecosystem. *Indian Journal of Traditional Knowledge* **7**(4): 642-654.

Chapter – 15

Highly Effective Fishing Methods in Central Assam as Perceived by Fishers

Rajita Devi and Biswarup Saha

INTRODUCTION

Assam is a heartland of indigenous groups. Since hundreds of years, many indigenous groups have been living in the state, maintaining their originalities in every sphere of life with difficulties. The tribal-indigenous groups constitute almost 12.4% of the total population of Assam. (Sentinelassam.com). Fishing is one of the major livelihood option of many indigenous fishers as the state is blessed with huge water resources with about 3.91 lakh hectare of water area in the form of rivers, *beel*, derelict water bodies and ponds and tanks. Besides swamps and low-lying areas, two major rivers and their tributaries have a high growth potential for fishing and could play a pivotal role in the socio-economic development and employment generation in the state. Indigenous fishers have developed a close and unique connection with the water resources and its environment in which they are depended for generation to generation for their livelihood. They have generated unique system of knowledge, innovation and skills relating to the uses and management of fishing activities. Over centuries, people have learnt how to get food and to survive in different environment. Farmers are very keen to observe the problems arising during fishing and seek solutions to adjust to their local environmental conditions. They continuously adapt and manage their fishing systems and in this way, they develop a rich storehouse of practical knowledge on their situation.

The traditional practices are considered to be as old as domestication of fishing practices. These practices have been percolating from one generation down to the next by oral transmission and considered to be the holistic approach for aquaculture management methodologies adopted by non-literate cultures. Historically, traditional knowledge was a part of problem solving for the local people and has used by the local people for management of their resources on a sustainable basis (Khan, 2000). Several experiments by the farmers, on trial and error basis have been taking place in the field of fishing over thousands of years. They made a wise use of available natural resources and devised many technologies to increase the quality and quantity of the output. The knowledge thus generated over the years is time tested and has the attribute of eco-friendliness. Such knowledge is called the 'indigenous technical knowledge (ITK)' or 'local knowledge' or 'traditional knowledge'. This knowledge is based on experience, often tested over centuries of use, adapted to local culture and environment and is dynamic and changing.

Recently, some ITK is lost naturally as practices get modified or are left unused for long time periods, the current rate of loss can be attributed to modernization and cultural homogenization, the current educational systems that believe macro-level problems can only be addressed through the global knowledge pool, and the slow growth of institution supporting grassroots innovations. A lot of traditional knowledge is also being lost because of the communication gap, since neither children nor adults spend as much time in their communities. Because ITK is generally transmitted orally, it is susceptible to change, particularly when people move to new regions, or when people's lifestyles tend to be different from that of their ancestors. On the above mentioned background the ex post facto study was undertaken to explore the logics of prevalent traditional fishing practices among fishers of Nagaon district, Central Assam.

METHODOLOGY

The study was conducted in the Nagoan district of Central Assam. Out of the 18 blocks in the district of Assam four blocks were selected with the help of purposive sampling method based on traditional fishers' populations. Further, two villages from each block were selected purposively based on the population of traditional fishers. From each village 6 active fishers who have the experience around 10 years were selected through snow ball sampling technique. Thus a total of 48 fishers were selected. The ITKs were documented through interaction, observation and focus group discussion with farmers by the help of semi structure interview schedule. Rationality refers to the degree to which ITK can be explained or supported with scientific explanations, or have been established based on long term experiences (Husain and Sundaramari, 2011). Rationality scale (Hiranand, 1979) was used to judge the rationality of the indigenous practices. After collecting the indigenous practices from the farmers, the practices were listed and send to a panel of thirty judges i.e., fishery scientists of Raha Fishery College, AAU, Assam and College of Fisheries, CAU, Lembucherra, Tripura with 4-8 years experience in fishery science for judging the rationality of the documented ITKs and their response were rated on 5 point score. The practices which were assigned a weighted mean score above 3.5 considered to be rationale. Perceived effectiveness of the ITK, i.e., the degree of relative usefulness of the ITK as perceived by the farmers in fishing activities, was measured using the Perceived Effectiveness Index (PEI) methodology (Sundaramari, 2001). The expert's opinion was also considered along with the perception of the farmers/ fishers practicing the respective ITKs to measure the MPEI.

RESULT AND DISCUSSION

A total of twenty six ITKs were documented on indigenous fishing gear, trap and methods and were administered to thirty experts on this field to evaluate their rationality. Five ITKs were identified as not rationale or unsustainable practices which should not be promoted. Then, the remaining 21 selected ITK was administered to same experts for analyzing their effectiveness on the line of catch per unit time by considering cost effectiveness, economy of time, economy of labour, simplicity, sustainability and local compatibility. The fishers perception in this regards were also collected. Of these, high mean PEI values were obtained by 4 traditional practices and declared as rationale and highly effective. These are 1. Gill net (*Langi jal*) 2. Bheta di maas dhora 3. Katal fishing 4. Drag net (MPEI score is 2.87, 2.68, 2.75 and 2.54 respectively). The highly effective ITKs are described below with scientific rationality.

1. Gill net (*Langi jal*)

These are rectangular gill nets fitted with head rope and foot rope (Fig.1a and 1b). The head rope is attached to floats and foot rope is made thicker with heavy ropes. The net is mostly operated during monsoon season (May to September). Usually the net is fastened to bamboo poles or allowed to drift overnight. Three types of Langi jal was found in the study area based on the targeted fish species. *Puthi langi jal* is used to catch *Puntius* species (locally known as *Puthi*); whereas *kawai langi jal* and *goroi langi jal* are used to catch *Anabas testudineus* (locally known as *Kawai*) and *Channa punctatus* (locally known as *Goroi*) respectively. They are long wall of webbing hang, set vertically in water. In this method the fish enter the craft voluntarily but it is prevented from coming out by means of certain arrangement. It is generally used to catch small and medium sized fishes. Mild water current is beneficial for its operation. This gear is perceived as simple and effective by majority of the respondents.

Key informants and practicing fisher: Mr. Akhtar Ali, Vill-Sutarpur, Block-Binakanti (Fig.1a)

Experts' comments: These nets are operated throughout the year in rivers, *beels*, paddy fields and derelict water bodies. The catch is miscellaneous and

Fig.1a: Gill net (*Langi jal*)

Fig. 1b: Gill net (*Langi jal*) in operation

the size of the webbing varies on its utility and the area of operation. Also some of the gill nets are designed to catch a specific kind of fish such as *puthi langi, goroi langi* etc. This net is also widely seen among the people of Assam due to low making cost, ease of operation. The gear is very effective for catching fish for daily consumption.

2. *Bheta di maas dhora*

This practice is done mainly in the post monsoon season. In this technique Fishers encircled suitable place with the help of mud and aquatic weeds (Figure 2). Then the water is removed inside the area with the help of buckets. When water level reduces people start catching fish by hand picking. This technique is also used by some fishers in small rivers and canals during monsoon season. Fishers select a suitable place where width of the river is less. Then they try to block the river with the help of tree branches, weeds etc. When fish will face the obstacle in the free flowing river they will jump to avoid it and fall on the *bheta*. At the same time people waiting in search of fish will actively catch the fish.

Key informants and practicing fisher: Mr. Feroz Khan, Vill-Murajar, Block-Binakanti

Fig. 2: *Bheta di maas dhora*

Experts' comments: *'Bheta diya'* is a barrier or a bamboo constricted screen set across a river, stream or channels of *beels*. The barrier guides the fish into the passage at the mid of the screen and the fishes are caught by a lever-operated dip net, installed at the end of the passage. In certain cases, the dip net is absent and the passage ends into a raised bamboo made platform. The method is used to catch all types of fishes during winter season when then the water current is less. This trap is used for commercial fisheries and involves high cost for its installation.

3. *Katal* fishing

Katal are erected just after monsoon (August- September) when the water level of *beels* start receding and fish is harvested during the month of January-February. *Katal* are round in shape having diameter ranging in between 20-35 meters (Fig. 3). *Katal* are set by dumping branches of plant twigs, bushes etc., in a depth varying from 1'5-2'5 meters and a circle made by fixing tree stumps around this vegetation mass to avoid scattering. The quantity of plant materials needed for the fabrication of the *katal* depends on the size and water hyacinth is barricaded with bamboo pole and strips. Fish are harvested during January–February using dragnet, cast net and local boat. Few days before harvesting, fishers encircle the *katal* with bananas and dragnet to prevent the fish sheltered below from escaping. Water hyacinth and tree branches are then removed manually using boats and reduce the encircled area by dragging the net. It facilitates easy harvesting. *Katal* act as fish aggregating device as the fish of different size and species find shelter in this structure for the purpose of feeding, breeding or to escape from predators etc.

Key information and practicing fisher: Mr. Jamal Miya, Vill-Roumari, Block-Batadrawa

Fig. 3: *Katal* fishing

Experts' comments: This kind of fishing is based on the principle of FAD. This fishing technique is operated in the oxbow lakes of river Brahmaputra and its tributaries, floodplains and reservoirs. The technique is used for commercial purpose and the fishes caught are sold in the market. The *beel* mohaldars/lessees usually install a *katal* in their leased *beels* to get the catch with ease. The operation is labour intensive and involves 4-12 persons depending on the size of the '*katal*'. The catch is miscellaneous. This method is indigenous and is practised in the state for a long time. The method is very effective in large water bodies where certain nets and gears cannot be operated.

4. Drag net

It is a small triangular drag net having mesh size of 45-50 mm. The net is fixed with bamboo frame for making it suitable for fishing operation. Usually fishers drag it for certain distance through the shallow portions of the river and *beel,* after which it is quickly lifted to catch the fishes in the middle portion of the net. Main seasons for the operation of this net are post monsoon and winter.

Key informant and practicing fisher: Mr. Sadik, Vill-Tuktuki, Block-Batadrawa

Fig. 4: Drag net

Experts' comments: These are large sheet of webbing used to surround a certain mass of water and drag towards the shore/banks or to a boat. These net is extensively used in rivers and on commercial reasons. Large sized fishes are its usual catch. The gear is labour intensive and requires several persons to operate with a boat or a few boats. The cost of gear is also expensive but can be compensated with the prized catch. The net is usually treated with some preservatives to increase the shelf life. The gear is mostly operated by the skilled fishermen communities of the state.

CONCLUSION

The present study documented 26 traditional practices on fishing. Majority of the ITK analyzed for their rationality were adjudged as rational by experts. Such rational and effective ITK suited to the local situations and local culture may either be suggested for adoption or may be recommended to the scientists for further examination or blended with modern fish production technologies, which in turn would promote sustainable fisheries system.

ACKNOWLEDGEMENTS

This chapter forms part of the MFSc thesis submitted to College of Fisheries, Central Agricultural University, Lembucherra, Tripura by the first author. The authors express their sense of gratitude to the Dean and other faculty members of College of Fisheries, CAU, Lembucherra and faculty members of Raha Fishery College, AAU, Assam. The authors are also grateful to the respondents (scientists and farmers) of the study.

References & Recommended Readings

Hiranand, 1979. Technocultural profile of a Dryland Village and Dry Farming Technology — An International Study, Unpublished Ph.D. Thesis, Dept. of Extension, HAU, Hisar.

http://www.sentinelassam.com/op_ed/story.php?sec=33&subsec=0&id=107532&dtP=2012-02-24&ppr=1 retrived on 21.04.12.

Husain, S. and Sundaramari, A. M., 2011. Scientific Rationality and Perceived Effectiveness of Indigenous Technical Knowledge on Coconut (Cocos nucifera L.) Cultivation in Kerala, *Journal of Tropical Agriculture* **49**(1-2), 78-87.

Khan, M. S. H., 2000. Attitude of Proshika Beneficiaries towards its Social Forestry Programmes. M. S. (Ag. Ext. Ed.) Thesis, Department of Agricultural Extension Education, Bangladesh Agricultural University, Mymensingh.

Sundaramari, M., 2001. Adoption and perceived effectiveness of indigenous agricultural practices in different farming systems. PhD thesis, Gandhigram Rural Institute, Gandhigram, India, 247.

Chapter – 16

Nu Ta: An Indigenous Fish Catching Gear of Arunachal Pradesh, India

Pema Chotten Wangja, Olik Jomang, Krishna Prasad Sharma and Yumlembam Jackie Singh

INTRODUCTION

Northeast India comprises the contiguous seven sister states.viz., Arunachal Pradesh, Assam, Manipur, Meghalaya, Mizoram, Nagaland and Tripura and the Himalayan state of Sikkim. The Northeastern states of India are blessed with diversified fish fauna which assumes special interest due to the fact that the region is grained by two important drainages *viz*., Brahamaputra-Barak and Chindwin Irrawaddy. (Gurumayum and Choudhury, 2009). Riverine fisheries resources of Northeastern States comprise of 19,150 km of streams and rivers. Except the Brahmaputra and Barak Valleys in Assam and Imphal Valley in Manipur, the Northeast mainly comprises of Hills and Mountains which form 65% of its territory (Suguman, 1998).

Arunachal Pradesh which falls under the Eastern Himalayan Region: one of the global mega biodiversity centres, is a home of 110 ethenic communities (tribes) most of which are forest dwellers and so diverse that they cannot understand each other's language (Srivastava, 2008). Community Fishing is a part of the cultural heritage of the tribal communities of North East which are yet to be documented except for that of Assam.

NU-TA OF ARUNACHAL PRADESH

Nu-ta is an indigenous passive fishing gear used to catch riverine fishes in few places of the West Kameng district of Arunachal Pradesh *viz., Rupa, Sher-gaon, Ji-gaon,* etc. It accounts for 8.86% of the total area of the state. It lies approximately between 91° 30' to 92° 40' East longitudes and 26° 54' to 28° 01' North latitudes.

'Nu' refers to fish and *'ta'* refers to the gear. The people in this region are mostly interested in fishing and hunting activities. Fishing is done for domestic consumption, sale at local markets and also for recreation. Nu-ta is used to catch fishes in small streams as well as in large rivers. It looks like a bridge in appearance. The main body with the structure of the bridge is called the *Nu-Ta*.

CONSTRUCTION AND MECHANISM

Nu-ta is constructed by diverting the flow of water in rivers. It requires more than one person to construct the gear. It is made up by a group comprising of twenty to thirty person and the number might vary depending on the width of the river and also on the size of the gear to be constructed. A favourable site is required for the construction of this fishing gear. *Nu-ta* is made up of bamboos, leaves and ropes. The following are the requirements for the construction of the gear:

1. A suitable site: A favourable site is required for the construction of this fishing gear. The gear has to be constructed in an area which has a good bank in order to create a pathway for the diverted river. Also the diverted river has to be connected back to the main river at another point.

2. Favourable season: The season used to catch fish by *nu-ta* ranges from May to August. This is so because the rest of the months are colder and hence the fishes do not swim much and rather prefer staying under the rocks and hence the catch is not high.

3. Manpower: The people constructing and using the gear comprise of around 30-35 people. These groups of people work together during the construction of the gear, but when it comes to collecting the catches, they divide themselves into different groups and collect in different batches.

4. Materials required are bamboos, ropes to tie the up the bamboos, stones to make up the pillars or the stands to hold the gear and leaves which helps in diverting the water by filing up the gaps that cover the water.

5. Temporary house: The fishers also construct a temporary house made up of bamboos and leaves in order to stay at night to catch the fish; a fire is also made; so that the fishes caught can be instantly dried. A group of fishers stay overnight and on the next day, another group stays overnight. This process is repeated for around two to three months bringing about catches every day.

After a suitable site for constructing the gear is selected; a small diversion of the river is created at the side of the river so that the water gets divided into two pathways. This diversion is done by digging up and removing the stones in order to let the river flow. On the new pathway *Nu Ta* is constructed. Long bamboos are placed together comprising of 8 to 10 bamboos of 5 meter length and giving the gear a width of half meter. The bamboos are tied together with the ropes. On the sideways certain bamboos are fixed perpendicular to the width of about 20 cm in height. This blocks the fishes from falling out from the sideways. *Nu ta* is constructed with its head in the diverted water so that the water flows right in to it and the tail of the *nu-ta* goes out of the diverted water into the bank. The *nu-ta* is constructed in such a way so as to oppose the current of the water. This makes the gear work similarly as a sieve functions in making tea. During tea preparation the sieve is held in such a way that it opposes the flow of the tea leaves along with the tea and as the tea passes through the sieve only the liquid part comes out and the tea leaves are stuck in the sieve. Similarly in this case, the water is allowed to flow through the gear whereas the fishes get stuck in the gear i.e. the *nu-ta.*

The water flows to the head of the *Nu-ta* but does not go all the way to the tail of the nu-ta as it falls down in the pathway of the diverted water, due to perforations and gaps between the bamboos. The diverted river then goes further and meets the main water body.

After constructing the *nu-ta*, supports are constructed for the stability of the *nu-ta* so that it does not collapse from the strength of the water current. The supports are constructed by putting large stones together and piling them up on the side of the nu-ta. These stones are held together with the help of bamboos which are used to keep the stones erect.

MECHANISM

When the *nu-ta* is constructed and the pathways are completed the main water body is blocked so that most of the waters flow through the new pathway and crosses the *nu-ta*. The water is blocked with the help of stones and leaves to form an obstruction in the main river pathway, thus the water is bound to flow through the new pathway. It is to be noted that some water still passes through the main route but the majority of the water is diverted towards the new pathway. When the river reaches the head of the *nu-ta,* the water flows through the *nu-ta* but as the fishes are large they cannot pass through it and instead get stuck in the bridge like gear with no water and is pushed out to the bank of the water. Hence, these fishes are caught easily by the fishers. The fishes caught are of large size as the small fishes manage to escape through the gaps. But if the fishers intend to catch the small size fishes a small bamboo mesh is placed beneath the nu-ta. Thus the smaller fishes going through the larger gaps are stuck in the smaller meshes bamboos.

Principle

Sieving is the main principle of the gear. There are two cases in which the gear uses the current to catch the fishes (Fig. 1).

Fig. 1: *Nu Ta* constructed in the river

1. The fishes go upstream for breeding or some other reason during monsoon season. On returning i.e. after monsoon when the fishes come downstream after breeding the fishes are sieved by the gear. The gear is made against the current of the water in the opposite direction to the water flow, so the current of the water pushes the fish towards the gear and hence are sieved in the gear.
2. Also the other fact is that fishes cannot go upstream in heavier downward current. During monsoon as the water level is high, fish finds it easier to go upstream but as the water level decreases and when fishes come downstream they are unable to fight the water current going towards the gear which is made with its mouth against the water current and hence are sieved. Also during monsoon as the river size increases gear is not made as it leads to the destruction of the gear.

Fishes caught

The fishes caught ranges from 5 kg to 20 kg per day. Fishes are mainly used for personal consumption though sometimes the fishers do sell the fishes.

Species caught

The species mainly caught are-'Nu chhlu'- loaches and 'nu sau'- *Schizothorax* sp.

Remark

Nu-ta is a highly efficient fishing gear, being one of the best riverine fishing gears in the locality. The drawbacks are it requires man power and is used seasonally.

DISCUSSION

Information on Indigenous Fishing methods of Arunachal Pradesh is very scanty owing to difficult terrain and remoteness of this predominantly hilly state (Dutta and Dutta, 2013). With diverse resources of biodiversity, people of different tribes in the state are practising different indigenous techniques to catch fishes. However, a very few of these practices have been documented and the need of the hour is to give more emphasis on this area. Community fishing is a part of the cultural heritage of the tribal communities of the North Eastern India which are yet to be documented except indigenous knowledge related to fish harvesting of North East region of India.

References and Recommended Readings

Dutta, R. and Dutta, A., 2013. *Bheta* Fishing- A traditional fishing practice of *Nocte* tribe of Tirap district, Arunachal Pradesh. *Indian Journal of Traditional Knowledge* **12**(1): 162-165.

Gurumayum, S. D. and Choudhury, M., 2009. Fishing methods in the rivers of Northeast India. *Indian Journal of Traditional Knowledge* **8**(1): 237-241.

Suguman V. V., 1998. Fishery resources of the Northeast Region and scope for their development, *Fishing Chimes* **18**(1):64.

Srivastava, R. C., 2008. Traditional knowledge of *Adi* tribes of Arunachal Pradesh on plants. *Indian Journal of Traditional Knowledge* **8**(2): 146-153.

Chapter – 17

Shifting Responses of Tadpole Larvae and Frog From Fish Pond in Presence of Hypoxia, Obnoxious Gases and Predators

Ramanuj Chakraborty, Subrata Shib and Himadri Saha

INTRODUCTION

Fish is a dependable source of animal protein in developing countries like India. In recent years, with the rapid development of the aquaculture enterprises, the infectious diseases caused by fungus, bacteria, virus and parasites became more severe, resulting in huge economic losses. Due to certain unavoidable problems for diagnosing and treating diseases, losses are being born by the poor fish farmers of the country.

The issue is more relevant to rural area like north-east India, where soil and water are highly acidic due to the high content of ferrous ions and high rainfall and their management needs high inputs. The situation gets shoddier when the farmers do not get the access to water and soil laboratory to test their pond samples. It is very hard for the rural farmers to monitor their culture pond condition without any scientific tools and technical knowledge. Moreover, stresses due to over-crowding, wide temperature fluctuations, cold stress, global warming, microorganism, pesticides, insecticides, chemicals, contaminants etc. are also incur more losses to the aquaculture production. These stresses result in heavy mortality, less growth and immune suppression etc. However, traditional fish farmers manage their pond and reduce above stresses by utilizing their technical knowledge which they gain through their experience as well as from their forefathers. In this chapter we are going to discus about a traditional knowledge which is being utilized by the farmers of rural Tripura to monitor and correct the health condition of their culture ponds.

METHODOLOGY

The information on the indigenous knowledge in fisheries, titled, "Shifting responses of tadpole larvae and frog from fish pond in presence to hypoxia, obnoxious gases and predators" was perceived as one of the feed-back received from the group of farmers after conducting a 5 days training programme entitled as "Post stocking management in aquaculture" on 10th December, 2012 at Salema R. D. Block, Tripura. The respondent group (trainee group) was consisted of 16 numbers of trainees leaded by Shri Ratan Lal Namasudra, a progressive farmer from Salema block, Tripura. After completion of training programme a farmer-KVK (Krishi Vigyan Kendra) specialist interaction session was held in connection with getting feed-back as well as to find the best possible available solution for their problems in pisciculture. Some of the farmers raised a critical point on shifting responses of tadpole larvae and frog during the culture period after heavy loading of organic and inorganic manure in their pond. The mentioned point was dully acknowledged to the farmers with a group discussion after the valedictory session of the training programme. The very next day a group discussion was made with the farmers and subsequent visit was made to the farmer's pond, where the practice is being carried out (Fig. 1).

Fig. 1: Culture pond where the farmer practice the ITK

RESULTS AND DISCUSSION

Rural poor farmers usually do not have the access to water and soil quality testing laboratories for monitoring and managing their ponds. However, in the aquatic environment, all aquatic organisms were exposed to a wide variety of environmental stressors. Fish and vertebrates respond to a stressor by eliciting a generalized physiological stress response, which is characterized by an increase in stress hormones secretion and consequent changes that help to maintain the animal's normal or homeostatic state (Barton, 2002) and they can also maintain their homeostasis my avoiding the stressors i.e., by escaping. A detailed experience sharing with the farmers reveal that the tadpole larvae shifted due to formation of gas in the pond (Fig. 1). They also expressed the thought and belief on the mentioned fact from their forefathers. The forefathers of the respondent advised orally to get possible remedial measures on removal of obnoxious gases from pond after looking the shifting behavior of tadpole larvae and frog.

The hypothesized belief of the respondent's forefather may be due to several logical rather scientific factors as mentioned here: Frogs typically lay their eggs in water. The eggs hatch into aquatic larvae, called tadpoles, like other amphibians, oxygen can pass through their highly permeable skins (Burton, 1972). For the skin to serve as a respiratory organ, it must remain moist. Ammonia is a form of nitrogen found in organic materials and many fertilizers. Toxic gases like Hydrogen sulfide (H_2S) can form in pond sediments when bacteria feed on organic debris in areas that are low or depleted of oxygen, giving off a rotten egg smell when the sediments are stirred up. The oxygen level in pond also decreases due to several other factors. This makes frogs susceptible to various substances they may encounter in the environment, which are toxic and can dissolve in the water film and be passed into their bloodstream. Moreover the carnivorous predators in the pond feed on the tadpole as well as the frogs, will also influence their abundance in the ecosystem. The frog's moist skin allows gases to pass directly between the blood vessels and the outside. When toxic materials from chemical fertilizers and pesticides flood the freshwater ecosystem, frogs will be among the first vertebrates to die. That is why many environmental scientists also believe amphibians, including frogs, are good biological indicators of broader ecosystem health because of their intermediate position in food chains, their permeable skins, and typically biphasic lives (aquatic larvae and terrestrial adults) (Phillips, 1994).

Thus frogs can be used as 'indicator species' for monitoring health condition of the pond and the shifting response of tadpole larvae and frog from fish pond indicate the presence to hypoxia, obnoxious gases and predators. In response to these proper remedial measures need to be applied to avoid fish kill. Those farmers are using another traditional technique to mitigate the problem. They apply banana leaf/stem ash to the pond after noticing shifting responses of tadpole larvae and frog from fish pond in presence to hypoxia, obnoxious gases and predators. Banana leaf/stem was ignited and formed the ash which was applied to their pond. The real quantity of ash not known but 1-2 banana trees per 0.16 ha were taken for the said purpose. Present article is expected to provide some basic information for monitoring and managing pond in traditional way and this can be adopted in other places, however the technology need to be validated with scientific research in near future.

References & Recommended Readings

Barton, B. A., 2002. Stress in fishes: a diversity of responses with particular reference to changes in circulating corticosteroids. *Int. Comp. Biol.* **42**: 517-525.

Phillips, K., 1994. *Tracking the Vanishing Frogs*. Penguin Books.

Burton, M., 1972. *The Observer's Book of British Wild Animals*. Frederick Warne & Co. pp. 204–209.

Chapter – 18

Application of Indigenous Technical Knowledge for Treating Ectoparasitic Infestations in Animals

W. Ramdas Singh, Debasish Behera, Hijam S. Devi, and P. H. Patra

INTRODUCTION

The medicinal values of plants are known since prehistoric times. Interestingly, 13000 BC old drawing on Lascaux caves in France suggests that humans were using plants for medicinal purposes. Ancient Egyptians chewed willow bark to relieve fever and headaches; and thousands of years later, scientists discovered that the bark contains salicylic acid- the active ingredient in aspirin. History reveals that in most of the countries natural products including plants, animals and minerals have been the basis of treatment of diseases both in human and veterinary medicines.

Plants as a source of medicine has been an ancient practice and is an important component of the health care system in India. The health management of people and domestic animals is based partially on the exploitation of natural resources, which needs proper documentation and research.

India - a developing country- whose economy is largely based on agriculture; where farmers derive their income from agriculture, livestock farming, fisheries, etc. The farmers face significant economic loss when their livestock suffer from diseases leading to loss in their productivity.

Ectoparasitic infestations in animals is one of the most common health problems in livestock , but they are not given due consideration and are usually neglected by the farmers; reasons being erratic supply and prohibitive cost of veterinary drugs and their supplies, poor communication facilities in many remote and hilly areas, and a shortage of manpower. For the small farmers with low income, these infestations can lead to enormous and insidious losses in the herds by weakening the body of animals and by transmitting diseases to other healthy animals of the herd which are often difficult to diagnose. Though, with the development of modern medicine, traditional veterinary practice have been replaced by, and overlooked at the international level, the use of plants and herbs in treatment of diseases, especially in animals and birds, should always be encouraged as they are easily accessible, cost effective and comparatively less toxic.

On the other hand, the synthetic pesticides such as organophosphates, organochlorines, synthetic pyrithroids, carbamates, etc. eventhough are effective against many ectoparasites in animals adversely affect the human health indirectly by consuming the toxic pesticides through animal products like milk, meat, eggs and their products which are contaminated with the pesticides residues. Most of the pesticides, due to their persistence and lipophilicity tend to accumulate in lipid rich tissues and get biomagnified. Moreover, farmers, in general, being unaware of the importance of withdrawal period, hardly observe it before slaughter or sale of milk or eggs for consumption.

Seeing the constrains in the veterinary medicine it is utmost necessary to better understand and enhance the knowledge of herbal remedies used in the management of ectoparasite infestations in animals. In due course, people will be able to explore certain indigenous technical knowledge regarding the application of traditional veterinary remedies used by local farmers in the treatment of ectoparasite infestations in livestock and other animals. These practices have been accumulated over generations living in a particular environment and are percolated from generation to generation through oral transmission. Understandably, these valuable practices are not properly documented and hence risk extinction before proper validation.

Some of the documents regarding the application of indigenous technical knowledge (ITK) in the treatment of ectoparasite infestation in animals are mentioned here:

1. *Azadirachta indica*

Common name: Neem

Family: Maliaceae

Neem is an evergreen tree native to India and Burma. It is a fast growing tree that can reach a height of 15-20 m, rarely to 35-40 m. Its leaves are used against ticks, flies, lices and insects. The leaves are boiled in water for 10 minutes and the resultant decoction is used to bathe the animals; it cures all forms of ectoparasitic infestations. The interval of bathing depends on the severity of the infestation. Crushed and mashed fresh leaves are also applied directly on the affected portion of the skin.

Fig. 1: *Azadirachta indica*

2. *Nicotiana tabacum*

Common name: Tobacco, Tambaku (Hindi)

Family: Solaceae

Tobacco, a native of Tropical America is cultivated as the principal source of tobacco. Tobacco contains the alkaloid nicotine, a powerful neurotoxin that is particularly harmful to insects. Crude powder of tobacco is applied on the animals' body and is proved to be effective against mites and lice. Decoction or crude extract in water is also used to bathe the animals.

Fig. 2: *N. tabacum*

3. *Mimosa pudica*

Common name: Touch-me-not, Sensitive Plant; Chui-mui, Lajwanti (Hindi)

Family: Mimosaceae

The plant grows wild throughout the tropical world. Its native origin is unclear. Its fern-like leaves close up and droop when touched, usually re-opening within minutes.

It has prickly stems and small, fluffy, ball shaped pink flowers in summer. The bathing of animals with the decoction (for about 15 min) of leaves and roots are effective against scabies and other allergic conditions.

Fig. 3: *Mimosa pudica*

Fig. 4: *Gliricidia sepium*

4. *Gliricidia sepium*

Common name: Gliricidia, Quick-stick, Mexican lilac

Family: Fabaceae

This tree is native to seasonal dry forest areas of Mexico and Central America. Small to medium-sized, thornless, leguminous tree up to 10-12 m high. Branching frequently from the base with basal diameters reaching 50-70 cm. Bark is smooth, varying in colour from whitish grey to deep red-brown. The leaves are boiled in water for a few minutes and the resultant extract is used as insect repellent and also applied on the skin to remove dermatophytes.

Fig. 5: *Millettia extensa*

5. *Millettia extensa*

Common name: Agarbel, Gauj, Gonj, Kauj and Patani (Hindi)

Family: Fabaceae

A shrub or a woody climber common in forests having yellowish brown bark. The leaves are pinnately compound. The whitish or brownish flowers are borne in axillary clusters. The powdered roots of this shrub is sprayed on the body coat to protect animals from ticks, lice and other pests.

Fig. 6: *G. sesquipedalis*

6. *Goniothalamus sesquipedalis*

Family: Annonaceae

The leaves of this shrub are used as insecticide. The leaves are crushed properly and rubbed on the affected areas of animal's body.

7. *Saussurea lappa/Saussurea costa*

Common name: Saw-wort and Snow lotus

Family: Asteraceae

Fig. 7: *Saussurea costa*

This plant is native to cool temperate and arctic regions of Asia, Europe, and North America. The plant is perennial herbaceous ranging in height from dwarf alpine species 5–10 cm tall, to tall thistle-like plants up to 3 m height. The leaves are produced in a dense basal rosette, and then spirally up the flowering stem. The flowers form in a dense head of small capitula, often surrounded by dense white to purple woolly hairs; the individual florets are also white to purple. The roots are strongly scented and yield aromatic oil which is used in making insecticides. The powdered dried roots are also sprayed on the animal's body to protect from various ectoparsites.

8. *Bombax ceiba*

Fig. 8: *Bombax ceiba*

Common name: Silk cotton tree, Cotton tree or Tree cotton; Semal (Hindi)

Family: Bombacaceae

This tropical tree grows to an average of 20 meters and has a straight tall trunk which bears spikes. Its leaves are deciduous in winter. Red flowers with 5 petals appear in the spring before the new foliage. It produces a capsule which, when ripe, contains white fibres like cotton. The bark, fruit and flower are used for treatment of various skin diseases including ectopasitic infestation in animals.

9. *Acanthus mollis*

Fig. 9: *Acanthus mollis*

Common name: Bear's Breeches

Family: Acanthaceae

This plant is native to the Mediterranean region from Portugal and northwest Africa to east Croatia, and is one of the earliest cultivated species of garden plants. It grows to 2 m tall, with basal clusters of deeply lobed and cut, shining dark green leaves up to 1 m long and 20 cm broad. The flowers are tubular, whitish, lilac or rose with spiny green or purplish bracts, and produced on stout spikes which grow up to 2.5 m (8 ft) above the leaves. It flowers in late spring or early summer. It grows in dry areas, and is tolerant of drought and shade. The leaf is crushed and made into paste and applied on the animal's body to kill lice.

10. *Curcuma longa*

Common name: Turmeric; Haldi (Hindi)

Family: Zingiberaceae

Curcuma longa L. is native to Asia; widely cultivated in West Bengal and other parts of India. This herb is with large ovoid rootstock, sessile tubers thick, cylindric, bright yellow inside. Rhizome paste is directly applied on the ectoparasite infested areas of the skin and continued till the animal is fully cured from the infestation. The paste also acts as a repellent against houseflies.

Fig. 10: *Curcuma longa*

Besides these, there are several other plants and herbs which are used in the treatment of ectoparasite infestations and other skin ailments in animals especially the livestock; some of them are- Agar wood (*Aquilaria mallaccensis*), Calumba wood (*Coscinium fenestratum*), Toon (*Toona Ciliata*), Blue ginger (*Alpinia galanga)*, Deodar cedar, Himalayan cedar, or deodar (*Cedrus deodara*), Black Sage or Wild Sage (*Cordia curassavica*), Bitter melon, Bitter gourd or Bitter squash (*Momordica charantia),* Arhar, Pigeon Pea, Red gram (*Cajanus cajan*), False Daisy, Bhringraj (*Eclipta prostrate*), Lipstick tree, Annatto, Arnatto (*Bixa orellana*).

CONCLUSION

The traditional knowledge inherited by the indigenous people of different regions in India is vanishing very fast. The traditional remedies used by the people for the treatment of diseases in animals and birds are being gradually replaced by more toxic synthetic drugs and chemicals available in the market, which are less cost effective, more toxic and not easily accessible to the farmers in remote villages; and also are human health hazard through the residues present in milk, meat, eggs and their products. Therefore, these precious indigenous knowledge which have been transferred from generation to generation verbally need to be properly explored and documented systematically. Effort need to be made to make aware the farmers, especially in remote and hilly areas, of the medicinal properties of the plants/herbs available in their regions and to encourage them to apply such remedies in the treatment of some of the diseases of their animals.

References & Recommended Readings

Bala, P., 1982. The Ayurvedic and Unani Systems of Medicine in Medieval India. *Studies in the History of Medicine* **6**(4): 282-289.

Costa, E. M. 2000. Sustainable development and traditional knowledge: A case study in a Brazilian fisherman's community. *Sustainable Development* **8**(2): 89.

Dutta, B. K. and Dutta, P. K., 2006. *Indian Journal of Traditional Knowledge* **4**: 7-14

http://www.flowersofindia.net.

http://www.manipurgovt.in.

http://www.theplantlist.org.

Jain, S. K. and Dam, N., 1979. Some Ethnobotanical notes from north-east India, *Economic Botany* **33**: 52-56.

Lal, J., Sambasivarao, K., Chandra, S., Naithani, R. C., Chattopadhyay, S. K. and Sabir. M.,1976. Studies on the comparative efficacy of *Cerdrus deodara* oil, benzyl benzoate and tetraethylthiuram mono sulphide against sarcoptic mange in sheep. *Indian Veterinary Journal* **53**: 543-545.

Lewington, A., 1993. A review of the Import of Medicinal Plants and Plant Extracts into Europe: Conservation and Recommendations for action, TRAFFIC International, Cambridge.

Mao, A. A., Hynniewta, T. M. and Sanjappa, M. 2009. Plant wealth of Northeast India with reference to ethnobotany. *Indian Journal of Traditional Knowledge* **8**(1): 96-103.

Mazid, M., Khanb, T. A. and Mohammada, F., 2012. Medicinal Plants of Rural India: A Review of Use by Indian Folks. *Indo Global Journal of Pharmaceutical Sciences* **2**(3): 286-304.

Pennington, T. D. and Styles. B. T., 1975. A generic monograph of the Meliaceae. *Blumea* **22**: 419-540.

Prakasha, H. M., Krishnappa, M., Krishnamurthy, Y. L., Poornima S. V., 2010. Folk medicine of NR Pura Taluk in Chikamaglur district of Karnatka. *Indian Journal of Traditional Knowledge* **9**(1): 55-60.

Raja, C., De, B., Devanna, N. and Sen. S. 2012. North- East India an ethnic storehouse of unexplored medicinal plants. *Journal of Natural Product and Plant Resources* **2**(1):143-152.

Sanglakpam, P., Mathur, R. R. and Pandey. A. K., 2012. Ethobotany of Chothe tribes of Bishnupur district (Manipur). *Indian Journal of Natural Products and Resources* **3**(3): 414-425.

Sinha, S. C., 1996. Medicinal plants of Manipur, Manipur Association for Science and Society, Imphal, India, p79.

Sullip Kumar M., 2008. Indigenous technical knowledge for control of insect pest and livestock disorders. *Indian Journal of Traditional Knowledge* **7**(3): 463-465.

World Health Organization. Guidelines for the Assessment of Herbal Medicine Programme on Traditional Medicine. Doc. WHO/TRM/91.4.WHO, Geneva, 2003.

Chapter – 19

Traditional Fermented Fish Products of North Eastern States of India

Yumlembam Jackie Singh, Huirem Bharati and Pema Chotten Wangja

INTRODUCTION

The North Eastern Region of India comprises of eight states, *viz*., Assam, Arunachal Pradesh, Manipur, Meghalaya, Tripura, Nagaland, Mizoram and Sikkim. These north eastern states are home to myriad tribals and several ethnic groups. Most of the people of the north-eastern region of India belong to the Mongolian race and the food habit of the region is fairly similar with that of South-East Asian countries. Fermented foods form an intrinsic part of the diet of these people. Fermentation has been an age long traditional practice of preserving various food products in the region. These fermented products are used as condiments in the preparation of various traditional sauces, strews and dishes (Saisithi, 1994).

Some of the popular indigenous fermented fish products are *Ngari* and *Hentak* of Manipur, *Seedal* of Assam and Tripura, *Tungtap* of Meghalaya, *Suko ka machha*, *Gnugchi*, *Sidra* and *Sukuti* of Sikkim, etc. (Jeyaram et al., 2009). Despite of some differences, most of these fermented fish products are being processed in a more or less similar way. The fermented products of the region are unique and exhibits peculiar taste and flavour that may not be favourable to everyone. This traditional art of preserving fish and fishery products by fermentation has been transferring from generation to generation in the north-eastern region of India.

NGARI

Ngari is an indigenous fermented fish product, traditional to the native people of Manipur. This fermented product is widely consumed as an intrinsic part of the daily diet by the local people. Attributed by its unique taste, flavour and therapeutic properties, *Ngari* is a popular daily household ingredient for the people of Manipur.

This fermented product forms an indispensable ingredient in most traditional dishes and curries as it attributes a unique ethnic taste, and the *Manipuris* (people of Manipur) very fondly consume it as a delicacy. It is consumed as side dish by the local people with cooked rice. *Ironba* is a local traditional dish prepared from vegetables, potatoes and chillies, where *Ngari* forms an inseparable ingredient.

PREPARATION OF *NGARI*

Sun dried *Puntius* are collected and then stored in gunny bags such that different sizes and grades are mixed together. The fish are stored in dried form to use for fermentation.

Jars are conditioned by smearing oil, mostly mustard oil on their inner surface. Prior to fermentation, the dried fish are washed with water using porous bamboo baskets and then allowed to drain the excess water for one night. Next morning, the fish are pressed by legs using gunny bags, thereby removing excess water by breaking the heads and bones of the fishes. It is believed that the oil released from the fish head upon pressing leads to initiation of the fermentation process.

The pressed dry fish is packed inside the earthen jars tightly. After packing the pots are then sealed with polythene sheet, fish scales, mud, oil slurry and cow dung slurry, and these packed pots are kept in the dark at room temperature for about 6-12 months.

In other cases, the fermentation is done in a way that the conditioned jars are first buried in pits upto their necks. The fish is put into the pots in such manner that about 5-6 kg of fish are put initially. It is then pressed under feet until wet liquid is released from the fish. This is continued till the pots are fully filled with fish. Then, these pots are sealed with mud and are kept for maturation for 3-6 months

After the maturation, the top layer of fish is generally removed from the pots. This top layer is locally called *Phumai* and it is usually considered as low quality fermented fish product. The rest of the product in the pots after removing the *phumai* is used for human consumption, and it is the *Ngari* as the local people call it. *Ngari* has a shelf life of about 12- 18 months.

The production cost of one earthen pot of *Ngari* goes upto Rs 1200- 5000. The product is sold in the market at the price range of about Rs 100- 240 per kg. The selling price of *Ngari* differs in market depending on its quality (Singh and Singh, 2009).

Skill is required in the preparation of *ngari* and for this, skilled labours are hired. Production and preparation of *Ngari* is a risk taking but very profitable entrepreneurship and due to the high capital requirement for starting this business, nowadays, *Ngari* production is mostly confined to the rich in the Manipuri society.

HENTAK

Hentak is another indigenous fermented fish product in paste form. Like *Ngari,* it is consumed by the Manipuris especially the *Meiteis*, an ethnic group of Manipur, as a regular ingredient for many food items taken along with cooked rice. Besides having a unique taste and flavour of its own, this fermented paste product has several nutritional as well as medicinal values and thus *Hentak* has been a traditional and native product for the local people since ages.

PREPARATION OF *HENTAK*

Hentak is generally prepared indigenously. The sun dried fishes of *Esomus dandricus* are first crushed to powder. The petioles of *Alocasia macrorhiza* are cut into several pieces and washed with water and then exposed to sunlight for about an hour. Equal weights of both fish powder as well as the cut and washed petioles of *Alocasia macrorhiza* are crushed to make a paste. Out of this paste, small balls are prepared. These balls are then put in earthen pots and stored at room temperature for around two weeks for fermentation. After two weeks of fermentation these balls are ready to use and are preserved. On keeping for few months they get hardened and are again propounded into paste with little water and stored as balls for reserve food.

Hentak is a special favoured fermented food item native to the state of Manipur and such kind of fermented paste product is rare in other north eastern states of India. This is traditionally prepared by the women folk of the state.

SEEDAL

Seedal is another fermented indigenous fish product of north eastern India, very popular in Assam and Tripura. (Arman et al., 2003). It is also prepared from small fishes mainly *Puntius* sp. *Seedal* is very similar to *Ngari* of Manipur even in taste and preparation but some distinct differences are also there between them. There are comparative size differences of the fishes used for preparation.

Ngari is prepared from comparatively small sized fishes ranging from 5-10 cm. The texture of *Ngari* is softer than that of *Seedal* which has hard solid texture.

The fermentation for *Seedal* is also similar to that of *Ngari*. Oil coating is also given on the inner surface of the earthen jars (*mutkas*). The oil processed *mutkas* are buried under the ground so that half of the belly of the pot remains under ground. Then the dug out soil surrounding the pot's belly is gathered so that the pot is tightly fixed in the ground. The pot should stand vertical to the ground. Gunny bags are used to spread over the surrounding of the pot to avoid any soil contamination. The packing of the fishes is also done in the case of *Seedal* by putting a layer of fish into the pot and pressing them under feet until wet liquid is released from the fish. The process is continued till the pots are fully filled. Then, the packed fish are covered with cover paste. The sealed pots are stored for maturation for about 4-6 months (Armaan and Basu, 2003)

All the raw materials used for *seedal* preparation are almost the same with those of *Ngari*. *Seedal* is fondly eaten by the local people as a condiment in various recipes with cooked rice and it is much liked as a daily food ingredient in these places because of its taste and flavour that it attributes.

OTHER FERMENTED FISH PRODUCTS

Tungtap is another fermented fish product indigenous to the Meghalaya state. (Agrahar and Subbulakshmi, 2006). This fermented paste product is consumed by the Khasi tribes of the state (Thapa, 2002). No much record has been available about its preparation.

Besides the Khasi tribes of Meghalaya, the Naga tribes spread over Nagaland and other adjoining states of the north east also preserve fish products traditionally.

The fermentation is done by the women folk of Naga tribes. Fish ranging from small to large are used. In case of larger fishes, they are cut into smaller pieces. The fish are washed and put in bamboo and plugged with leaves and usually kept over the fireplace for fermentation. The fish get fermented within few days and are ready to use for consumption. The fermented fish product can be stored for about a month. It cannot be stored longer as it gets rotten very fast and becomes unpalatable. This fermented product of fish is mainly prepared by the Lotha tribe of the Nagas (Mao et al., 2007)

Another fermented product is prepared from crabs. The black species of crab is used for this. The crabs are washed, their hard appendages and entrails removed and grounded. It is then mixed properly with ground black til (*Sesamum orientale*) and wrapped in banana leaf (*Macaranga indica* or *Phrynium pubinerve*) and this is kept to ferment over the fireplace in the typical Naga kitchen for about a week. On opening the banana leaf, it gives a typical inherent smell and it is ready for use in cooking various traditional dishes. This product is prepared indigenously by several Naga tribes like the Lothas, the Maos, and the Angamis. The black crab is said to have an aroma that gives the product a unique good taste.

Fermentation of foods is an indigenous trend in most north eastern states of India. Among the fermented foods, the fermented fish products are indeed very fondly taken as vital food items in the region.

SCOPE OF THE FERMENTED FISH PRODUCTS

Each fermented product of the north eastern region is associated with a unique microflora that increases its nutritional properties. Fermentation enhances the nutritional quality of any food product by increasing amounts of vitamins and protein solubility and also by improving amino acid patterns. These fermented fish products are prepared at the household level mostly through indigenous practices. In Manipur, many people are engaged in the production and preparation of *Ngari* to earn a livelihood. *Ngari* and *Seedal* are more or less commercially prepared and produced in a large scale in their respective states. But still there is a huge gap to reach the large scale commercialization and industrialization of these traditional indigenous fermented fish products.

There is need of awareness of hygienic knowledge of production of these traditional fermented fish products. Development of value added products by selecting micro organisms, genetic improvement, process improvement, raw material improvement, etc. will lead to industrialization of these products. Development of quality traditional fermented fish products with clinically health promoting properties could be marketed at a lower price. This will in turn provide healthy impacts on the health and welfare of the people especially the low income group of north eastern states of India.

CONCLUSION

Traditional fermentation of various foods has been one of the oldest and most commonly practised methods to preserve the foods for longer period in the north eastern states of India. Fermentation of fish and fishery products thus has been practising since ancient times by the local people. These fermented fish products indicate that these ethnic people of the north east India were associated with advanced technology of preservation of foods since ages. The medicinal and microbial properties of the fermented products was studied and documented by many workers (Sarojnalini and Singh, 1988). The high degree of hydrophobicity by some lactic acid bacteria (LAB) strains isolated from *ngari, hentak* and *tungtap* indicates the potential of adhesion to gut epithelial cells of human intestine, advocating their 'probiotic' character (Thapa et al., 2004). Lactic acid bacteria are normal residents of the complex ecosystem of the gastrointestinal tract (Holzapfel et al., 1998). The microflora associated in fermentation largely contributes in enhancing the nutritional quality and medicinal values of the preserved food products. Thus, these fermented products are considered as good sources of various essential nutrients for human consumption.

References & Recommended Readings

Agrahar, M. D. and Subbulakshmi, G., 2006. Preparation techniques and nutritive value of fermented foods from the Khasi Tribes of Meghalaya. *Ecology of Food and Nutrition* **45(1)**:27-38.

Ashiho, A. M. and Odyuo, N., 2007. Traditional fermented foods of the Naga tribes of North eastern, India, *Indian Journal of Traditional Knowledge* **6**(1): 37-41.

Arman, U. M. and Basu, S., 2003. Seedal: An Indigenous Fermented Fishery Product of Northeast India, *Fishing Chimes* **7**:30-32.

Arman, U. M., Basu, S. and Nayak, B. B., 2003. *Seedal* – an indigenous fermented fish product from India. *Info Fish International* **6**: 49-51.

Holzapfel, W. H., Haberer, P., Snel, J., Schillinger, U. and Huisin't, V. J. H. J., 1998 Overview of gut flora and probiotics. *International Journal of Food Microbiology* **41:** 85-101.

Jeyaram K, Anand Singh Th, Romi W, Ranjita Devi A, Mohendro Singh W, Dayanidhi H, Rajmuhon N and Tamang J. P., 2009. Traditional fermented foods of Manipur. *Indian Journal of Traditional Knowledge* **8(1):** 115-121.

Saisithi, P., 1994. Traditional fermented fish, fish sauce production, Fisheries Processing (Biotechnological Applications) Ed-A.M. Martin, Chapman and Hall, London. 111-129.

Sarojnalini, C. and Singh, W. V., 1988. Composition and digestibility of fermented fish foods of Manipur. *Journal of Food Science and Technology* **25:** 349-351.

Singh, C.A. and Singh. Y. J., 2009. *Ngari,* An indigeneous fish product of Manipur, *Aqua International,* June 2009.

Thapa, N., 2002. *Studies on microbial diversity associated with some fish products of the Eastern Himalayas*. Ph.D. thesis, North Bengal University, India.

Thapa, N., Pal, J. and Tamang, J. P., 2004. Microbial diversity in ngari, hentak and tungtap, fermented fish products of North-East India. *World Journal of Microbiology & Biotechnology* **20:** 599-607.

Chapter – 20

Indigenous Technical Knowledge (ITK) in Animal Husbandry Practices

Avijit Haldar, Chintu Deb Barma and Rajkumari Sanjukta

INTRODUCTION

In the emerging global knowledge economy, a country's ability to build and mobilize knowledge capital is equally important for sustainable development like the physical and financial capital (World Bank, 1997). The basic component of any country's knowledge capital is its indigenous technical knowledge (ITK). ITK is the social capital of the local people, their main asset to invest in the struggle for survival, to produce food, to provide for shelter or to achieve control of their own lives. It is the unique, traditional, local knowledge existing within and developed around specific condition of men and women indigenous to a particular geographical area (Ghosh and Sahoo, 2011). It helps to increase the sustainability of development efforts because of its integration process with mutual learning and adaptation, which in turn contributes to the empowerment of local communities. Thus, ITK has now been recognized and accepted as a vital knowledge source (Chimaraoke et al., 2003). It is particularly true with respect to the health care systems. World Health Organization (WHO) has stated that 80 percent of the world's population depends on traditional medicine for its primary health care and traditional knowledge is indispensable for its survival. The reconsideration of traditional medicinal systems in the industrialized world and the fact that modern medicine is too expensive for many developing countries are the main reasons for the decision of the WHO in the 1970s to promote traditional medicated systems by checking scientifically the efficacy of the plants used in traditional medicine and identifying the principle components responsible for genuine therapeutic effects (Bizimana, 1997). An increased demand of natural products in form of drugs, foods, cosmetics etc. has been noticed in recent years. Thus, the importance of documenting ancient/indigenous/ traditional knowledge has recently gained a momentum all over the world including India. In USA a Centre of Indigenous Knowledge and Rural Development (CIKARD) has been established at IOWA University in 1987. USA has also taken lead by establishing the Office of Alternate Medicine (OAM) in 1992 under National Institute of Health (NIH) for research and development of indigenous medicines (Singh and Chauhan, 2010). In India, several non-government organizations [BAIF, Pune; Jagaran Vikas Kendra, Udaipur; ANTHRA, Pune; SALIHOTRA, M.P.; Appropriate Technology of India, Ahemdabad; Women's Organization for Rural Fevelopment (WORD), Ahemdabad and Grass Roots Innovation Augmentation Network (GIAN), Gujarat] are reported to be involved in documentation and validation of ITK system.

There is no universally accepted definition of Indigenous Technical Knowledge. Many authors have defined it in different ways. None of the definitions is essentially contradictory; they overlap in many aspects. Wang (1988) defined ITK as sum total knowledge and practices which are based on people's accumulative experiences in dealing with situations and problems in various aspects of life and such practices are special to a particular culture. According to Farrington and Martin (1988), ITK can be defined as the basis for knowledge, beliefs and customs which are internally consistent and logical to those holding them, but at odd with the objectively deduced findings of formal science, it has comparatively more influence on people than that of modern substitute. Warren (1991) stated that ITK is the local knowledge – knowledge that is unique to a given culture or society. ITK contrasts with the international knowledge system generated by universities, research institutions and private firms. It is the basis for local-level decision making in agriculture, health care, food preparation, education, natural-resource management, and a host of other activities in rural communities. The word 'indigenous' means "native born, originating or produced naturally in a country or in a specified area" and the word 'knowledge' means "assured belief, practical skill, that which is known, learning, enlightenment" (Reijntjes et al., 1992). As per IDRC Report (1993), it is a complex concept that reflects on even more complex set of empirical, intellectual, social and spiritual factors that constitute human culture. It refers to the integrated expression of collective values and customs that guide interaction among peoples and between people and nature. Throughout the centuries, societies evolved by learning from experience and transmitting knowledge to younger generations. Flavier (1995) defined ITK as the information base for a society, which facilitates communication and decision-making. Indigenous information systems are dynamic, and are continually influenced by internal creativity and experimentation as well as by contact with external systems. According to Pushpangadan et al. (2002), ITK is a community based functional knowledge system, developed, preserved and refined by generations of people through continuous interaction, observation and experimentation with their surrounding environment. It is a dynamic system, ever charming, adopting and adjusting to the local situations and has close links with the culture, civilization and religious practices of the communities.

ITK AND INDIA

India, the country of Rishi and Krishi, has a very rich heritage of traditional health control and several treatment systems (Ayurvedic, Unani, Homeopathy) that have been used for human and animals since time immemorial. These practices have been percolating from one generation to another by oral transmission, folklore, rituals, songs and even laws. India is one of the World's 12 mega-diversity countries accounting for eight per cent of global plant genetic resources and a higher share of micro-organisms, most of them undocumented (Bidwai, 1997). Human relationships with plants kingdom is over 10,000 year old. Due to such a long association it is very intense, vast and multifarious. In Indian subcontinent ITK about plants dates back to Harappan civilization but not much information is available, may be, mainly due to its urbanized nature which involves less interaction with plants and trees and script used by Harappans is difficult to decode. On the other hand, during early settled life (1000 BC to 600 BC) Aryans gave birth to ancient Indian literature which was orally handed over to next generation. Later this information got documented. This process of recording information preserved the glorious ITK in India related to various facets of Indian culture. Many works on identification, documentation and scientific validation of ITK in India has been taken up through various projects under the previously ICAR, NATP. The Indian Council of Agricultural Research has collected and recorded 595 veterinary traditions from different sources (Swarup and Patra, 2005). Indigenous knowledge is continuously evolving and built from and based on thousands of years of experience needs careful documentation and analysis for developing sustainable farming systems, otherwise of which would be lost, not regain at any cost in the future (Talawar and Singh, 1993).

ITK IN ANIMAL HUSBANDRY PRACTICES

ITK in animal husbandry is considered as old as domestication of various livestock species. Unfortunately, these practices, which are in vogue throughout rural India, are little documented and there is danger of extinction of this knowledge which needs to be documented. This could be achieved through group discussion, personal discussion, Rapid Rural Appraisals (RRA) etc. Thus, it is of paramount importance to collect and document these practices and to assess their validity. ITKs may be validated by QuIK (Quantifying Indigenous Knowledge) method (De Villiers, 1996). The basic principle of this method is to give emphasis on experience and knowledge of the farmers who know and understand the environment in which they farm over the years and thus it allows to collect and tabulate the answers of the farmers to many questions on particular ITK. QuIK methodology represents a rapid and relatively cheap way to obtain indigenous technical knowledge. In QuIK, matrix ranking is combined with an interview schedule to generate numerical data from the experienced farmers. The matrix is designed through preliminary discussions with the farmers and used to obtain quantitative data. The experienced respondents are asked to weigh the ITK(s) in comparison with the modern veterinary practices for its performance on different criteria like effectiveness, accessibility, compatibility, adaptability, sustainability, rationality, limitations etc. The respondents may be asked to put required numbers of pieces of stone out of 10 in each block of matrix. The matrix results are the data for suitable analysis purposes.

Farmers are very rich source of ITK related to cattle rearing (Kaul et al., 1998). Different authors/researchers have documented different indigenous animal husbandry practices mainly concentrating on treatments of diseases. These are presented in a tabular form (Table 1).

Table 1: Indigenous technical knowledge in animal husbandry practices

Sl. No.	Indigenous knowledges	Uses	References
1.	Sulphur oil containing 65 g of gandhak (Sulphur) and 250 ml of Sarson oil	Skin disease	Khanna (1967)
2.	The tribal people of Chotanagpur use firing with red hotiron followed by washing of wound with boiled neem leaves	Pus formation in the swelling of yoke gall	Srivastava (1982)
3.	Application of a herbal plant 'Harjora' (*Vitis quadrangularis*) on the affected area and fixing it with dry bamboo splints	Fracture	Srivastava (1982)
4.	Oral feeding of three to four seeds of 'Bhilama' (*Semecarpus anacardium*)	Anoestrous in cattle	Gupta and Patel (1992)
5.	'Hangingof Thor' (*Euphorbia* sp.) branch at the gate of animal byre	The notion is that the animals do not get infection of food and mouth disease (FMD)	Mane (1992)
6.	Use of 'Kulpaimeni' (*Acalipaindica* sp.)	Respiratory disorders, dry cough and bronchitis	Venkatasu bramanian and Fulzele (1993)
7.	Feeding of 20- 25 numbers of 'jack fruit leaves' (*Artocarpus* sp.) with oil immediately after parturition	Expulsion of placenta	De (1994)
8.	Feeding of Kalajira mixed with sour whey	Intestinal worm	Khatik (1994)
9.	Bamboo leaves (*Bambusa* sp.)	Diarrhoea	Dohare (1996)
10.	Feeding of spider eggs and flour of ragi by Toda tribes of Nilgiris	FMD	
11.	Application of mohua oil collected after extraction from seeds of mahua (*Madhuca indica*) as used by the Santal tribes of West Bengal	Hoof injugy in FMD	Dana and Kaul (1998)
12.	(i) Drenching of fresh green leaves of Babla (*Acacia* sp.) and sugar, or (ii) boiled and mixed juices obtained from bark of Banjam (*Aridisia solanaceasyn*) and from Ban kurchi (*Holarrhena antidy sentrica*), or (iii) juice obtained from bulk of Sal tree (*Shorea robusta*)	Diarrhoea	Dana and Kaul (2000)
13.	Drenching of properly mixed of ashes of oyster shell with molasses	Tympanitis	Dana and Kaul (2000)
14.	Feeding of masoor dal (*Lensculinaris medic*) and kutcha halood	Retention of placenta	Dana and Kaul (2000)

(Contd.)

Sl. No.	Indigenous knowledges	Uses	References
	(*Curcuma lonaga* Lin) mixed with molasses by the Santal tribes of West Bengal		
15.	Feeding of vela seed (*Seme-carpus anacardium*) after proper grinding and mixing with mustard cake	Foot and Mouth Disease	Dana and Kaul (2000)
16.	Application of a paste prepared with kutcha halood (*Curcuma lonaga* Lin) and lime	Wound	Dana and Kaul (2000)
17.	Application of Bonroson (*Alliaria sativum*) and mustard oil after proper boiling	Wounds of mouth in contagious ecthyma	Dana and Kaul (2000)
18.	Drenching of (i) juice obtained from roots of Isharmul (*Aristolochia indica*) after proper mixing with warm water, (ii) juice obtained from bark of Chatim (*Euphorbia nerifoila*) after proper mixing with warm water	Respiratory problems	Dana and Kaul (2000)
19.	Seed of kernel of mango (*Mangifera indica*) after proper grinding and mixing with half- boiled rice as used by the Santal tribes of West Bengal	Weakness in poultry birds	Dana and Kaul (2000)
20.	(i) Drenching of 100 ml sap collected from the extract of Thankuni (Indian pennywort, *Centella asiatica*), Patharkuchi (*Coleus aromaticus*) and Durba (Dhub grass, *Cynodon dactylon*) for 2-3 days, (ii) Feeding of 3 pieces of Golmorich (Black pepper, *Piper nigrum*), 5 g Jowan (*Bishop's weed, Trachyspermum ammi*) and 5 g *Chirata* (Swertia angusti*folia var pulchella*) are grinding for 3-4 days, (iii) Feeding of 200 g Kalmegh (Creat, *Andrographis paniculata*) leaves and 100 g Thankuni leaves (Indian pennywort, *Centella asiatica*) after grinding. These are practiced by the tribes of Paschim Midnapore district of West Bengal.	Dysentery in cattle	De et al. (2004)

(*Contd.*)

Sl. No.	Indigenous knowledges	Uses	References
21.	(i) Feeding of 50 g Amlaki (Emblic myrobalan, *Emblica officinalis*), 50 g Haritaki (Chebulic myrobalan, *Terminalia chebula*) and 50 g Bahera (*Terminalia bellirica*) once daily for 7 days,(ii) Feeding of 10 g bark of Aswatha (Banyon, *Fiscus benghalensis*), 10 g Ada (Ginger, *Zingiber officinale*) and 10 g salt once daily for 7 days.	Bloat	De et al. (2004)
22.	(i) Application of hot fomentation of a paste made with Akanda leaves (Asclepiadaceae, *Calolropis gigantea*) and ghee, (ii) Application of a luke worm paste prepared from Rasun (Garlic, *Allium sativum*) and ghee.	Arthritis	De et al. (2004)
23.	(i) Feeding of 7 pieces of chicken egg per day for 7 days, (ii) Feeding of 12 pieces of Kala (Edible banana, *Musa paradisiaca*) along with 400 g sugar for 2 days, (iii) Drenching of 200 g bark of Asok tree (Ashoka, *Saraca asoca)* in 1 litre boiled milk once daily for 15-20 days.	Anoestrus in cattle	De et al. (2004)
24.	Feeding of a paste made up of the extract from the roots of Bonson tree and 21 pieces of Golmorich (Black pepper, *Piper nigrum*)	Dog bite	De et al. (2004)
25.	Feeding of a viscous suspension consisting of 50 g of til bark or leaves, 50 g of turmeric, 50 g of lal duddhi, 50 g of sarson grains, 100 g of tulsi leaves and 100 ml of lemon juice @ 50 g twice a day for 2 days in Bundelkhand.	Haemorrhagic septicaemia (HS) in cattle	Singh and Chauhan (2010)

'ITK' IN ANIMAL HUSBANDRY PRACTICES AVAILABLE IN NORTH EAST INDIA

Indian North Eastern (NE) region ranks 6th among top 25 biodiversity spots in the world. The NE region of India is a region of magnificent natural, cultural multitude and consists of rich flora and fauna. The NE part of India lies between 21.5^{0}N to 29.5^{0}N latitude and 85.5^0 E to 97.5^0 E longitudes. The region falls under high rainfall zone and the climate ranges from subtropical to Alpine. The region is characterized by difficult terrain, wide variations in slopes, altitude, land tenure systems and diverse cultivation practices. The NE region today constitutes of about 7.8% of total land space of the country. The uniqueness of NE India lies in the fact that each state possesses a wide range of tribes with diversified cultures, traditions and dialects. More than 200 numbers of tribal populations contributes to a unique cults and cultures. Tribal people are the repository of accumulated knowledge passed on by word of mouth from generation to generation and thus, permanently record their experiences and knowledge in a form inaccessible to outsiders. Therefore as tribal groups disappear, their vast knowledge vanishes with them (World Bank, 1982). The livestock rearing plays an important role in economic activities as well as in all tribal cultures, not only as a sole source of milk and meat, but also in the traditional rituals where the animals were found to be featured in almost all of their major ceremonies, festivals, rituals, dances and pantomimes. The livestock species commonly reared in these regions are cattle, goat, sheep, yak, mithun, pig, ducks, poultry etc. The overall economic development of the tribes of these regions depends on the productivity of these animal species due to extensive agricultural crop production being practically impossible. These regions have a very rich heritage of traditional methods for feeding, care and management, health control, treatment and processing techniques that have been used for animal husbandry practices. Some of the ITKs related to animal husbandry practices available in the north eastern states of India have been collected based on semi-structured questionnaire and personal interview. These are mentioned herein.

1. Animal breeds are mostly indigenous covering non-descriptive cattle, buffaloes, pigs (Naga, Mali and Dom), goats (Black Bengal, Assam Hill goats) and poultry birds. People still prefer to rear local breeds of animal species due to less incidence of diseases, less requirement of external feeds etc. This is also indigenous knowledge towards animal farming.
2. Mithun and yak play an important role in Nagaland and Arunachal Pradesh. These animals are the symbol of social status of an individual. These are also used for gift and dowry purposes in marriages. They are even worshipped by the tribes in social cultural system.

Fig. 1 Mali pig in Tripura

Fig. 2: Yak in Arunachal Pradesh

3. In Arunachal Pradesh, the yak herdsmen use male yaks for breeding the local hilly cow in order to produce cattle x yak cross. In the F1 generation, if the progeny is a male, it is called as Dzo, which is found to be sterile, whereas if it's a female, it is called as Dzomo that is fertile and produces double the quantity of milk than that of yak. However, Dzomo of F2 generation has been reported to be poor yielder. Hence, the yak farmers prefer to use male yak for the production of Dzomo of F1 generation. Presently, researches are being mainly carried out research on the Dzo and Dzomo especially at National Research Centre (NRC) on yaks to unfold the scientific facts on the production of Dzomo (Debbarma et al., 2009).

4. Animal shelter is made up of locally available materials like bamboo, cut woods, straw etc. These shelters appear primitive, but they are economical and sustainable. Elevated type shelter is also available to keep goats and poultry birds. This type of shelter has some advantages for easy cleaning facility and protecting animals from dampness in such high rain fall area. The system of rearing of pigs is primitive. Pigs are kept in a small enclosure made up of bamboo, woods or tin or in a small shelter or tying with a rope under a tree. This type of pig rearing system needs to be addressed to explore the advantages or disadvantages on socio-economical, human health and hygiene point of view.

Fig. 3: Local cattle shed

Fig. 4: Pigs are being kept under the tree

Fig. 5: Local pig sty

Fig. 6: Goat house

5. Mithun and Yaks are reared in semi domesticated conditions and they are reared in free range as the animals graze on the forest pastures. Though Mithuns mostly graze in the forest, interestingly they never fail to identify their owner's residence and reach home to meet up their thirst for salt. This is way of claiming the ownership of the animals. This is a common understanding among the tribes. This concept is purely local (Debbarma et al., 2008).

6. Bamboo made different types of baskets are very popular to keep and carry small animals and poultry birds. Bamboo made net is used to cover the mouth of bullock for preventing bullock to eat during farm operation. Bamboo splint is also used to join the fracture. Bamboo based indigenous technologies have come up, because of availability of different types of bamboo in the entire NE region.

Fig. 7: Bamboo made basket for carrying chicks

Fig. 8: Bamboo made basket for carrying piglets

7. In Tripura and other states, pigs are the traditional livelihood option for the tribal families. There is an unique method of fattening pigs. Rice bran mixed with locally made *Choak* (rice bear) waste is mainly fed to the pigs. *Choak* is country liquor brewed from the rice after fermentation using *Muli* made up of locally available certain plant leaves and roots. This *Choak* is consumed in most of the tribal families and the left over waste is available to be fed to the pigs. This probably results the restriction of movement of the pigs due to hallucination. *Choak* (rice bear) waste being rich in carbohydrate may favor quick fattening in those pigs. In some places, dried pork is grinded with rice grain and fed to the pigs to protect from cold and enhance growth. All these indigenous knowledge need to be validated for their importance in the farming system.

Fig. 9: *Muli* made up of locally available certain plant leaves and roots

Fig. 10: Rice bran mixed with locally made *Choak* (rice bear) waste is being fed to the pigs

8. In yak rearing practices, the Monpa tribe has been observed to feed tree leaves in addition to the grazing. Chatterjee et al. (2003) studied on some locally available tree fodder species commonly fed to yaks in Arunachal Pradesh and concluded that those plants possess some medicinal value as depicted in Table 2.

Table 2: Indigenous plants with anti-ailment properties in Arunachal Pradesh

Sl. No.	Botanical name	Local name	Anti-ailment properties
1.	*Acer canbelli*	Syluli	Back ache, Cataract, Conjunctivitis, Dysentery.
2.	*Ligustrum myyrsinites*	Zimbu	Antibacterial, Astringent, Immunomodulator, Laxative.
3.	*Costanpsis* spp.	Maar	Increases milk yield
4.	*Quercus* spp.	Phrengpa	Dysentry, Diarrhea, Nasal cattarrah, hemorrhoids
5.	*Salyx humboltiana*	Salyx	Rheumatism, Epilepsy, Piles and Swelling

Various indigenous herbal medicines used by the Yak keeper especially in Arunachal Pradesh has also been documented by Ramesha et al. (2009) (Table 3).

Table 3: Indigenous medicinal palnts of highlanders of Yak tracts

Sl. No.	Indigenous medicines	Local name	Uses
1.	*Thalictrum foliosum*	Nyan-thub	Inflammation and fever
2.	*Berbaris* sp.	Kangchang	Diarrhoea
3.	*Barbaris* sp.	Ser-shing	Conjunctivitis, sore mouth, infection of throat and larynx
4.	*Aralia* sp.	Main-sela	Diptheria, leprosy, constipation, body ache, joint pain
5.	*Rubus ellipticus*	Kanda-kari	Cold and infectious fever
6.	*Dryopteris* sp.	Dum-bu	Poisoning
7.	*Rubus idaeus*	Rasatika	Fever, common cold, cough, lung disease
8.	*Aconitum heterophyllum*	Shin-tu-dug-med	Antiseptic, antidote against snake, scorpion bite, fever
9.	*Drynaria* sp.	Ngur-mo	Poisoning
10.	*Hedychium spicatum*	Sunithera	Fibrinolytic agent
11.	*Picrorhiza kurrao*	Khonglong	Diarrhoea
12.	*Nardostachys jatamansi*	Pang-pos	Fever, spleen disorder and swelling
13.	*Mirabalis himalaica*	Ba-spru	Dropsy, debility, reduces excess accumulation of fluid in the joints
14.	*Saussurea costus*	Ruta	Anodyne, antibacterial, carminative
15.	*Pterocephalus hookeri*	Karta	Cough, asthma, colic pain, gout and arthritis

9. In some places, bamboo flowers and leaves are fed to cattle in view of early ovulation and less chances of retention of placenta.
10. Hentak (fish product), snail and shell are fed to poultry birds for more egg production and bigger size of egg.
11. Guava leaves, banana stem juice are fed to cure dysentry.
12. Ant hill soil in water is boiled and then pasted on the foot of cattle after cooling and extract of the boiled ant hill soil is put on the mouth to cure foot and mouth disease in cattle.
13. Fermented fish mixed with turmeric and garlic is fed to the poultry birds as preventive measure for protecting the poultry birds from any disease and also increasing appetite.
14. The NE region is full of flora and fauna. A lot of local plants are used as indigenous medicines in veterinary practices by the tribal people. However, the efficacy, side effects, cost effectiveness of those indigenous medicines warrant thorough investigation for further developmental and promotional purposes. A few indigenous medicines used in veterinary practices are listed in Table 4.

Table 4: Indigenous medicines used in veterinary practices

Sl. No.	Indigenous medicines	Local name	Uses
1.	*Allium porrum* Linn.	Maroinakupi (Manipuri)	Fever
2.	*Tinospora cordifolia*	Ningthou-khongli (Manipuri)	Foot and mouth disease in cattle
3.	*Adhatoda vasica* Nees.	Nongmangkha (Manipuri)	Cold, cough and fever
4.	*Andrographis paniculata* Wall ex Nee, leaves	-	Cold and fever
5.	*Plectranthus ternifolia* D. Don.	Khoiju (Manipuri)	Mosquito repellant
6.	Leaf extract of *Tera paibi, Micrania micantha, Tridax procumbens*	-	Wound healing
7.	Paste of mustard oil and camphor	-	Relieve joint pain

15. Crushed maize, salt and *Zanthoxylum nepalense* berries are mixed with yak blood to prepare sausages encased in intestines and finally boiled to cook for human consumption (Ramesha et al., 2009).
16. A traditional method of processing of milk and production of Churpi (country cheese) in Arunachal Pradesh is depicted in Flow Chart 1.

Flow Chart 1

Collection of milk
↓
Heating up to 30-40°C
↓
Churning
↓
Separation of cream
↓
Curdling of the milk
↓
Churning of the curd
↓
Separation of Ghee
↓
Heating of the left out portion (skim milk)
↓
Addition of the culture/starter with the skim milk
↓
Separation of soft churpi (country cheese)

17. A traditional method of processing of milk and production of Cheese Candy/ Churcum (Dry Cheese) in Sikkim is presented in Flow Chart 2.

Flow Chart 2

collection of the fresh country cheese
↓
Wrapping of the fresh country cheese in square shaped jute made gunny bag piece
↓
Compression of the wrapped cheese very tightly with a flat stone for 3- 4 hours
↓
Unwrapping of the square shaped cheese
↓
Sun drying
↓
Churcum/ Cheese candy/ Dry cheese

CONCLUSION

Today, many indigenous knowledge (IK) systems are at risk of becoming extinct because of rapidly changing natural environments and fast pacing economic, political, and cultural changes on a global scale. The World Bank states that IK systems are "at risk of becoming extinct" (World bank, 1999) so there is a need to search for more effective and creative interactions between indigenous knowledge and scientific knowledge systems (Dewalt, 1994). While many ITKs are in a danger of extinction, many ITKs have not been documented or realized particularly available in NE region. ITKs may be useful based on the cost, accessibility, compatibility, adaptability and sustainability. However, these may be sometimes less effective than the modern veterinary practices (De et al., 2004). On the other hand, ITK is compatible with the local situation and easy to adapt, being often less dependent on the use of external inputs. For example, the traditional practice of applying turmeric and mustard/ coconut oil to wounds is very easy to adopt, whereas modern veterinary practice involves the use of external inputs like antibiotic preparations and sometimes the services of a stockman or a veterinarian to treat the wounds. Thus, it is necessary to identify the perceived effectiveness as well as sustainability of such indigenous practices for their further scientific validation in various national and international research organizations. Studies of traditional wisdom can not only benefit social scientists, but can also act as eye openers for the biological scientists (Kaul, 1994). ITK is unique to a given culture and society, but it has value also for the scientists and planners that are involved in development projects (Rao et al., 1995). Indigenous knowledge systems can help in the adoption of technology packages where there is poor cooperation with farmers (Lightfoot, 1995). It may provide problem solving strategies for local communities, especially for the poor. ITK may allow us to unpack the practices of successful farmers, so that information can be disseminated to a wider group of farmers. Each technology has limitations and ITK is no exception. The indigenous materials and methods used vary from area to area, because ITK is unique to a particular place. So, the scope of popularization of ITK may be addressed keeping in view the geographical locations. Nevertheless, there is a need to screen the ITKs, to cull the non-sense and to recommend the relevant ones. What is the truth that feeding excreta of birds can induce heat, that application of mud on wounds can help the healing, that removing of testicles through open method by a butcher or a cobbler is better than the castration etc. On the contrary, some age old practices have some real scientific meanings. For example, many dairy farmers in Indian villages milk the first few strippings on the floor, even though they do not know the scientific logic that fore milk has maximum bacterial load which can reduce the keeping quality of milk if mixed with the clean milk. However, in organized herds, this milk is used for mastitis testing by stripping the fore milk from all the quarters on to strip cup. The integration of appropriate IK systems into development programs may bring sustainable development impact on farming community. The best example is the use of many herbal medicines in combination with allopathic medicines in the treatment of livestock diseases in all parts of our country. Finding an optimum mix and balance between indigenous participation and scientific participation in research and development processes rather than a choice of either one or the other is required in finding best innovations. As IK is the information base for a society that facilitates communication and decision-making, it needs to be well documented and disseminated to the corresponding target communities to achieve better empowerment which leads to development of the nation. It is necessary to develop databases of success stories, best practice lessons from projects and provide access to these databases and disseminate the same through publishing or electronic format.

References & Recommended Readings

Bidwai, P., 1997. Blocking biopiracy: We must not trip over TRIPS. The Times of India, March 12th 1997.

Bizimana, N., 1997. Scientific evidence of efficacy of medicinal plants for animal treatment, Ethno veterinary Medicine: Alternatives for Livestock Development, Proceedings of an International Conference held in Pune, 4-6 November, 2: Abstract. pp. 11-12.

Chatterjee, A., 2003. Feeding management of yaks. *North East Veterinarian* **3**(1):18.

Chimaraoke, O.I., Ugai, G.A., Ukwayi, J.K., 2003. Indigenous knowledge and communal conflict resolution: evidence from Nigeria. *African Journal of Indigenous Knowledge Systems* **2**(2).

Dana, S. S., Kaul, P. N., 1998. Indigenous veterinary medicines. *Honey Bee* **9** (4): 14.

Dana, S. S., Kaul, P. N., 2000. Indigenous technical knowledge in veterinary medicine among tribals. *Indian Journal of Animal Research* **34**: 56- 59.

De Villiers, K. A., 1996. Quantifying indigenous knowledge: A rapid method for assessing crop performance without field trials. Agricultural Research and Extension Network, Agren, July 1996.

De, A., Arya, H. P. S., Tudu, B., Goswami, A., 2004. Indigenous technical knowledge in animal husbandry. *Livestock Research for Rural Development* **16**(8):1- 10

De, H. K., 1994. Indentification and assessment of indigenous technologies in animal husbandry in Bankura district of West Bengal. M.Sc. Thesis, NDRI, Karnal.

Debbarman, C., Singh, B. P., 2009. Study on Socio-Cultural and Traditional Values associated with Yak rearing in Eastern Himalayan Regions. *Indian Journal of Social Research* **50**(4).

Debbarman, C., Singh, B.P., Khan, M.A., 2008). Migratory pattern of Yak Herdsmen in Eastern Himalayan region: A case of Arunachal Pradesh and Sikkim. *Livestock International* **12**:18-19.

Dewalt, B. R., 1994. Using indigenous knowledge to Improve agriculture and natural resource management. *Human Organisation* **53**:123-131.

Dohare, R.S., 1996. A study on indigenous technical knowledge in animal husbandry in athura district. Ph.D. thesis, IVRI, Izatnagar.

Farrington, J., Martin, A., 1988. Farmer participation in agricultural Research- a review of concepts and practice, Occasional paper No. 9 ODI.

Flavier, J. M., 1995. The regional program for the promotion of indigenous knowledge in Asia. In: Warren, D.M., L.J. Slikkerveer and D. Brokensha (eds), The cultural dimension of development: Indigenous knowledge systems. Intermediate Technology Publications, London. pp. 479-487.

Ghosh, P. K. and Sahoo, B., 2011. Indigenous traditional knowledge. *Orissa Review*. **1**: 65- 70.

Gupta, A. K. and Patel, K. K., 1992. Survey on farmer's innovations in Gujarat. *Honey Bee* **3** (2): 22.

IDRC Report,1993. Morales on knowledge change and the preservation of progress, by Daniel A. International Development Research Centre, *Canada* **21**: 4- 5.

Karthikeyan, C., Chandrakandan, K., 1996. Indigenous technical knowledge of the tribes in Agriculture. *Journal of Extension Education* **7**: 1417-1421.

Kaul, P. N., 1994. Traditional wisdom in veterinary science. *Farm Digest* **4**: 25-26.

Kaul, P. N., Mishra, O. P, Kumar. A., Lal, B., 1998. Cattle rearing practices of Bareilly farmers. Dairy Guide Jan-March, 1998.

Khanna, B. M., 1967. A study on the indigenous systems of Veterinary medicine as practiced by the farmers of Hissar-1 Block. M.Sc. thesis, Punjab Agricultural University, Hissar.

Khatik, G. L., 1994. A study on training needs of tribal dairy farmers in Rajasthan. Ph.D. thesis, NDRI, Karnal.

Lightfoot, C., 1995. Using indigenous knowledge systems in the design of on-farm experiments- a Philippine case. In: Warren, D. M., Slikkerveer, L. J., Brokensha, D., eds.), the cultural dimension of development: indigenous knowledge system. International Technology Publication Ltd., London, U.K. pp 348-353.

Mane, P. M., 1992. Branch of Euphorbia sp. *Honey Bee* **3**(2): 20.

Pushpangadan, P., Rajasekharan, S., George, V., 2002. Indigenous Knowledge and benefit Sharing- A TBGRI experiment in IK strategies for Kerala. NSE Publication, Thiruvananthapuram.

Ramesha, K.P, Bora, M., Kandeepan, G., Chakravarty P., 2009. Indigenous Traditional Knowledge of Yak rearers. Technical bulletein. Publised by the Director, NRC on Yak, Dirang, 790101, Arunachal Pradesh.

Rao, S. V. N., Van Den Ban, A. W.,Rangnekar, D. V., 1995. Indigenous technical knowledge and livestock. In: Singh, K., Schiere, J. B, eds.), Handbook for straw feeding system. Indian council of Agricultural Research, New Delhi, India. pp. 119- 128.

Reinjtjes, C., Haverkort, B., Bayer, A. W., 1992. Farming for future: an introduction to low external input and sustainable agriculture; ILEIA, Leusdan, *Netherlands*, **52:** 210-218.

Singh, S.K., Chauhan, J., 2010. Perceived effectiveness of ITK among livestock owners. *Indian Research Journal of Extension Education* **10**: 12-15.

Srivastava, R. M., 1982. Cattle in culture and economy of tribal Mundas of Bihar. Ph.D. Thesis, Kurukshetra University, Kurukshetra.

Swarup, D., Patra, R. C., 2005. Perspectives of Ethnoveterinary medicine in veterinary practices. In: Nair, M.N.B (ed), Proceedings of the National conference on contemporary relevance of Ethnoveterinary Medical Tradition of India, Bangalore, India.

Talawar, S., Singh, Y. P., 1993. In search of wisdom among the farmers of arid agriculture: documentation and analysis. Paper presented at National seminar on Indigenous Technologies for sustainable agriculture. March 23-25, 1993, held at IARI, New Delhi.

Venkatasubramanian, V., Fulzeele, R. M., 1993. Indigenous knowledge and practices in animal husbandry. Paper presented at National Seminar on Indigenous Technologies for sustainable agriculture. March 23-25, 1993 held at IARI, New Delhi.

Wang, G., 1988. Indigenous communication systems in research and development. *Journal of Extension Systems* **11**: 75-86.

Warren, D. M., 1991.Using Indigenous Knowledge in Agricultural Development. World Bank Discussion Paper No.127. Washington, D.C.

World Bank,1982. Report of World Bank on Tribal peoples and economic development-human ecological consideration. World Bank, Washington, D.C.

World Bank,1997. Knowledge and Skills for the Information Age, The First Meeting of the Mediterranean Development Forum; Mediterranean Development Forum, URL:http://www.worldbank.org/html/fpd/technet/mdf/objectiv.htm.

World Bank,1999. World development report 1998/1999: Knowledge for development. Retrieved April 30, 2005, from http: //www.worldbank.org/afr/ik/ikrept.pdf.

Chapter – 21

Traditional Fishing Practices in North Eastern Region of India

Anil Datt Upadhyay

INTRODUCTION

Fish are one of the main exploitable resources of the aquatic ecosystems that provide a cheap source of protein which help solving the problem nutritional security particularly NE region of the country. The rapid development of fishing technology in India during the recent years has paved the way for increased production availability of new synthetic materials, evolution of new designs of fishing crafts and gears and avoidance of ignorant harmful fishing techniques etc. It is to be taken into account that any developmental activity adopted should also be base rooted in the objective of uplifting the economy of the fishing communities. This implies that the improvisation of the existing tools and techniques and introduction of new ones to enhance fishing efficiency need a careful study of the traditional infra-technologies. Further selection of fishing methods and gear are influenced by various factors such as physiographic of the water body, nature of fish stock, characteristics of the material from which gear are fabricated and standard of living. Therefore, variation in application of fishing devices can be observed in different type of water bodies, which have characteristic of their own due to unique nature of the water resources of the region. The North Eastern Region of the Country is very rich in the indigenous knowledge and techniques of fishing. The fishing gears and crafts vary from one state to other states in terms of technical specification, materials and method of fishing. The fishing methods and fishing devices which are in use in the region are the result of continuous efforts and knowledge inherited through generations of fishing community. In this presentation a review, compilation and documentation of fishing devices reported from North East region have been made.

PAST STUDIES ON FISHING TECHNOLOGIES OF THE REGION

The detailed aspects of fishing traps and pots have been described by Slack-Smith (2001). Many researchers reported various fishing traps operated in Indian waters [Hornell (1938), Job and Pantulu (1953), Prabhu (1954), George (1971), Brandt (1984), Mohanrajan (1993), Nair (1993)]. Mitra et al. (1987) have described the traps used in upper and middle Hooghly estuary. Details of traps operated in north eastern India was reported by Sharma et al. (1993). Traps used in Khachodhara Beel of Assam reported by Sharma and Ahamed (1998). Some efforts to document the type of traps used in Assam has been carried out by Bhagawati and Kalita (1987), Nath and Dey (1989), Choudhury (1992), Choudhury et al. (1996), Sharma (2001) and Bhattacharjya et al. (2004), Gurumayum and Choudhury (2009) and Baruah et al. (2010). Detailed designs of traps of Assam and its operation have been reported by Pravin and Meenakumari (2008). A traditional fishing method of catfishes using duck meat as an attractant was studied in Assam by Dutta and Bhattacharjya (2009). Fish Diversity and Fishing Gears used in the Kulsi River of Assam, India was reported by Islam et al. (2013). The use of dugout canoes and different type of fishing gears in Lotak Lake was reported by Hora (1921), Choudhuri and Banerjee (1965), Singh (1983), Devi (1980), Mao (1991) and Singh et al. (1992). Marak et al. (1998) explained the operational details of the fishing techniques carried out by the Tribal's' in North-Eastern India with reference to Meghalaya and Manipur. The traditional fishing methods and fishing devices of Mizoram have been reported by Lalthanzara and Lalthanpuii (2009). The traditional knowledge associated with fish harvesting practices of War Khasi community of Meghalaya was studied by Tynsong and Tiwari (2008). An account of traditional fishing devices of Tripura reported by Upadhyay and Singh (2013). Traditional fishing methods in Central valley region of Manipur and Traditional fish aggregating wisdom of Manipur were studied and reported by Devi et al. (2013).

The detailed account of traditional fishing techniques and fishing devices of Tripura, Assam, Mizoram, and Manipur are presented in following sections:

1. TRIPURA

In Tripura, capture fisheries resources constitute 37 percent and however, these resources contributes only 5.50 percent of total fish production. The capture fisheries is important in state of due to livelihood of 22,373 fishermen families are depends on it. A wide array of fish catching device used in Tripura and the devices used in lentic and lotic water bodies had variations with regards to their fabrications and operations. Fish catching devices based on design, materials used, technical details and operational methods may be broadly categorized as I. Fishing gears with netting, II. Hooks and Spears, and III. Traps:

I. Fishing Gears with Netting

a. Gill nets (Fash jal)

Though gillnetting is banned in the state, still it is very common fishing technique in shallow and moving water bodies of Tripura. Both selective and non selective types of gill nets with different mess size were found in operation in the Tripura. The details of the gill nets are as follows-

i. ***Fash jal / Kanke jal/ Chat jal/ Current net*:** Gillnets are widely used fishing gears during the rainy season especially in sallow moving water bodies in Tripura. There are single walled nets with mesh size 2-12 cm (mostly 5.2 to 6.2 cm) and length of the gear varies from 10 m to 50 m depending upon the width and depth of water bodies. It has been observed that gillnets operated in Tripura were mostly made with polyamide monofilament. Head ropes of gear were made up with poly propylene. The small stones or gravels are used as sinker. Single man with small *kusha/ dingi* boats fixes the nets with bamboo poles across the flow of water. This net is operated during night time from evening to morning. After 4-6 hours, fishes are collected from the net. The target species of fish are *Puntius, Mystus (tengra), Glassogobuis, Loaches, Chanda*, Indian Major Carps and prawn.

b. Drag nets

i. ***Laitya jal***: It is a surrounding net which is mostly used to catch *Channa* spp. One end is hold by fishermen on shore and a boat speeds the net and encircles the areas. Its mesh size is 2.5 to 3.75 cm. It is used to catch *Gudusia chapra, Chanda nama* in Rudrasagar lake. It is made up of nylon.

ii. ***Fy-jal***: It is a surrounding net, operated by two boats. First the boat goes to the centre of lake and then they spread the net by carrying each boat each end of the net and encircling the area. The two ends are pulled up and catch the entire lot of fishes. Generally it is made up of very small mesh size for catching pelagic small size fish.

iii. ***Ber-jal*** **:** It is the most popular dragnet of Tripura. Mesh size varies from 0.5-13 mm. About 60 % of resources are exploited by this gear. The length of net varies from 25-150 meter. Its breadth is 4-25 meters. It is mainly used for catching of carps. It is operated from boats in deeper lakes. It is one type of surrounding gear mostly used in domestic ponds. During operation, the net is spread horizontally and vertically one side of tanks by covering the whole breadth of the tanks with float and sinkers. Sometimes it is also called as "*Gharra jal*".

iv. ***Tana-jal*:** It is two men operated drag net mesh size of the net varies from 1 to 8 mm. Mouth is kept open to 4.5meters and its breadth is 1.8m Mouth is kept open by bamboo sticks of 1.2m height. It is dragged in shallower water. Small and medium sized fishes are caught during daytime. It is made of mosquito net hence it also called "*Masari jal*".

c. Bag nets

i. ***Pelni jal:*** It is a scoop net, having a round or triangular frame. The mesh size of *pelni jal* varies from 4 to 5mm and bamboo strip used to form small square. The diameter of bamboo frame is 50-60 cm. It may be operated by boat or by walking. *Puntius* spp., *Amblypharyngodon mola* etc. are caught by it.

ii. ***Phaloon/ Khara jal/ Bel jal*:** It is a traditional boat operated dip net. The netting is hung to triangular frame. The two bamboo poles of length 3-5 meter were tied togather to form acute angle. The net is fastened on two poles of bamboo while at the free end, it is supported by rope. The mesh size of the *phaloon* gear is ranged from 5 mm to 8 mm. The fabrication cost of the gear is Rs. 1200/ and expected economic life is 3 years. In

Fig. 1 & 2: Fishing with *Phaloon jal* in Rudra Sagar Lake of Tripura

good season a fisherman with small *dingi* boat catches on an average about 7-10 kg of fish per day. During rainy season it is fixed in canals facing the current of water to collect fishes came through drifting.

iii. ***Chapila jal*:** It is operated from two boats. During spreading the net, the fishermen make sound by leg over boat. It attracts the fishes to aggregate and the bag like net and then pulled up and the fishes are harvested. The size of net is length 100-150 m and breadth 3 m. Mesh size 6-10 mm. It is used to catch *Gudusia chapra* mainly.

iv. ***Maiya / Aain-jal*:** It is mainly to catch catfishes specially *Mystus* spp. in rivers. Float has a bit lesser weight than sinkers. During rainy season, it is most frequently used.

d. Falling gears

i. ***Kuni jal /Uran jal:*** It is a cast net operated by one fishermen. The principle is only to throw the net to cover the fish. Lower edge is provided with series of sinkers. It may be cast from bank of the rivers, beels, and ponds or from boat. It is provided with a line to haul it.

Fig. 3: Fishermen Operating KuniJal for fishing

Fig. 4: Village women knitting Kunijal

In case of Rudrasagar lake of Tripura, Kunijal (cast net) was used by 45.37 percent fishermen, followed by *faloon jal* (21%), *ber jal* (10%), gill net (9.24%) and *chapila jal* (8.40%) as reported by Upadhyay (2013).

e. Lift net

i. ***Dharma jal:*** It is used in running water especially in rivers. Frame is made up of bomboo. Net is made up of cotton twine. Its mesh size is 1-3cm. It is also widely used in beels, Chera riverlets. Mainly used for catching weed fishes. The length of bamboo frames is 5-7 meter.

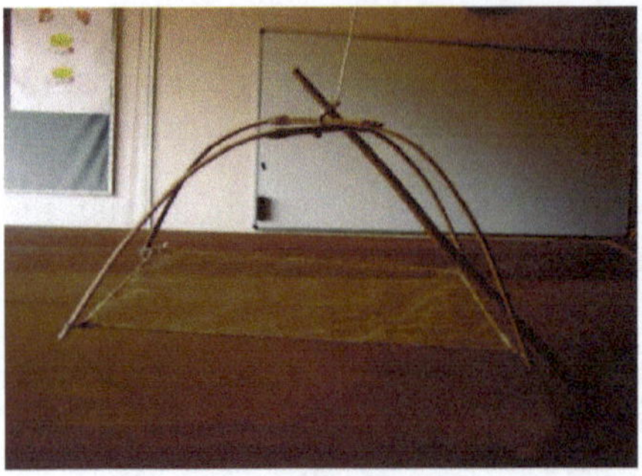

Fig. 5: Lift net

Fig. 6: Operation of *Dharma Jal* in river system

Fig. 7: Scoopnet

II. Spears and Hooks

Fishing hooks and lines are simplest and the most important among fish catching devices of Tripura. They are simple, easy to operate, cost effective and selective. The different types of lines and fishing hooks found in Tripura are described below:

a. Spears

i. ***Juitya*:** It is a long bamboo stick (70-80cm long). The tip of bamboo stick is equipped with 5-8 hooks and extreme end of stick is attached with a nylon rope. When the fishermen recognized that fish is passing nearby to him, he threw the "*Juitya*" that pierced to the fish. These are practiced by experienced or trained persons only. This practice used to catch mainly big size fishes like silver carp and some other carps.

ii. ***Gathynna/Guthia*:** Gathynna a steel or wooden stick with pointed end. This type of spears mainly used to locate *Monopterus cuchia* in swampy areas and marginal embankments. A fisherman first seal hole and then inserts the stick at soft ground randomly and when he feels the movement of *eel,* he tans out the *eel* by digging with hand.

iii. ***Ek-Kaithia***: The length of this spear varying from 1.5 to 2.5 meter. It has 4-10 sharp prongs made up by splitting seasoned bamboo pole at one end. Sometime this type of wounding spear referred as "*Chal*" which is used by the tribal fishermen in catching or fishing of large fishes. They pushed it in shallow areas and the fish cannot come out of it. Specially, *Cyprinus carpio* is caught by spearing with the above.

b. Hooks (Barshi)

"*Barshi* fishing" in Tripura is very common and old age practice. This type of fishing is specially used in shallow water bodies with weeds or standing crops and also to catch bottom fishes. It is operated throughout the year. A Nylon line is tied on the tip of bamboo pole. It is a type of hand line. The sinker in it is locally called as "Bhar". The hook is provided with lure/bait like earthworm, forage fishes, small prawns, bread and rice paste. The following types of *Barshi* are being used by fishing community to catch different type of fishes.

i. ***Boal Barshi***: A Nylon line is tied on the tip of bamboo pole. Length varies from 1-3 meters. The hook is provided with small live bait. As the name of the fishing hook, this used specially to catch reverine fish like boal fish (*Wallago attu*).

ii. ***Tang Barshi***: This type of hooks used in large scale during the flood. A Nylon line is tied on the bamboo stick. Length varies from 1-1.5 meters. The hook is provided earthworm, small prawns, bread, and rice paste. This type of fishing hook is used to catch *Puntius* spp., *Channa* spp. *and Tilapia* spp. fish.

iii. ***Gucha Barshi*:** This type of hooks used in large scale during the flood. A Nylon line is tied on the bamboo stick. The hook is specially designed and made up with wooden sticks. The hook is provided earthworm, small prawns, bread, and rice paste. This fishing hook, used to catch snake fish *Monopterus cuchia* generally found in the mud/soil of water bodies.

iv. ***Baira*:** This is an arc like hook made up with flexible bamboo stick. In the middle of the stick a Nylon line is tied and on other end line is tied with bamboo round stick of 9-12". The hook is specially used in either paddy field or weedy shallow water body to catch koi fish (*Anabus testudinous),* magur, singhi, kanla etc.

III Fishing Traps

i. ***Anta Trap:*** It is a rectangular box-shaped trap made up of bamboo wire mesh or iron or polythene strips. There is a small mouth opening which only opens in its inner side by water pressure. Earthworms, rice bran mixed with dry fish are given inside the box to attract prawns.

ii. ***Benta:*** This trap is like *Anta* with little difference that it has wide front and round or narrow back. It is also made up with bamboo strips and tightly woven with steel wire or plastic threads. It is fixed in moving water especially on the dyke of the paddy field. This trap is mainly used to catch fish like chanda, putti, baim and gutom etc.

Fig. 8

Fig. 9

Fig. 10

Fig. 11

Fig. 8-11: Different types of fishing trap made up of bamboo

iii. ***Singchai*:** It is tubular trap with circular mouth and tapered back portion of trap. It is made with narrow bamboo strips, bamboo sticks and woven with plastic threads. The *Singchai* is used to catch bottom feeder fish like magur, singhi, baim and baila, crab, prawn etc. in small shallow water bodies. It is fixed at the bottom, mouth opening across the flow of water.

iv. ***Pharam:*** It is a rectangular box-shaped trap which made up with bamboo strips. There is a small mouth with flexible opening which only opens in its inner side. Mouth opening is against current. It is set in stagnant weedy water body to catch medium and small size fish including IMCs. These are placed in a series marked with bamboo stick. These are placed in the evening hours. The fish entered in the trap but cannot come out from it. In the morning these fishes are collected

v. ***Deur*:** This trap is just like carry bag made up with bamboo strips, sticks and woven with plastic thread. This trap is set in the moving water to catch fish such as putti, baim, tengra, chanda and gutam etc.

Fig. 12: *Deur*

Fig. 13: *Pharam*

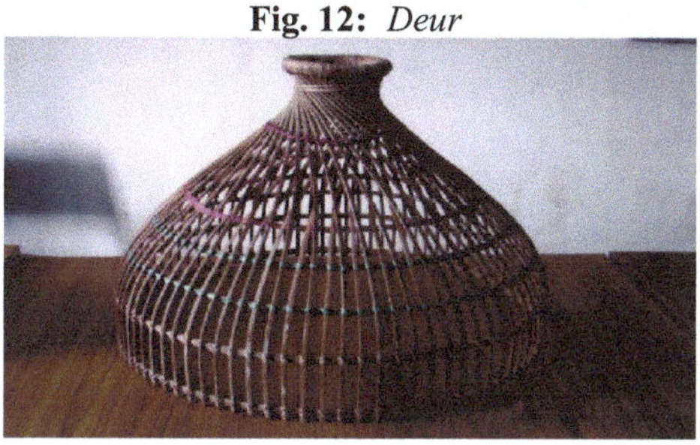

Fig. 14: *Phalo*

Fig. 15: *Singchai*

Fig. 12-15: Different types of fishing trap made up of bamboo

vi. ***Phalo:*** This trap is funnel shaped with circular bottom and narrow at upper side. The trap is made up with bamboo strips, sticks and woven with plastic or iron wires. This trap is very common in Bangladesh and some part of Tripura state. The small and medium size fishes were caught by this device.

vii. ***Jak:*** It is sometimes referred to fish aggregating device. It is mainly practiced in Rudra Sagar Lake as well as most of wild water body of Tripura. The fishermen keep bushes, *Eichhornia*, Bamboo poles with leaf in the water body. They also provide foods in it. Fish consider the "*Jak*" as their habitat, starts to congregated there and breed there. After 2-3 months the fishermen harvest the whole plot by mosquito net. They take out the bushes and are placed in a new location again. They harvest all the fishes residing at that "*Jak*".

Though Tripura is small state but indigenous knowledge and technology for fishing is very rich. In the study of Upadhyay and Sing, 2013 it was reported that The fishermen of the state were using nylon net for gillnetting which is harmful in case ghost fishing. They further emphasized on uses of bio-degradable netting materials for gill netting.

2. ASSAM

The fishing gears and craft in Assam are traditional, non- mechanized and mostly locally built. The highly diverse natural water areas, methods ranging from grapple with the operation of large and indigenously designed nets are adopted in Assam for capturing different varieties of fishes. Some of the common fishing gears of Assam are described below:

A. Fishing Nets Used in Assam

I. Encircling gear

The encircling gears are found to be operated in shallow water and the following types of encircling gears have been recorded:

i. ***Angthajal*:** It contains several pockets at the bottom end and is fitted with sinkers of iron rings. It is a cast net of larger sizes 3 to 4 m in length and 2 to 3 m circumference. This net is operated from boat and target fish are carps and cat fishes.

ii. ***Khewali jal*:** It is a type of cast net locally known as *Jhankijal* or *Asrajal*, is commonly used in Assam. This is a conical cast net with 8-15 mm mesh size. A rope is attached to the apex of the cone. The end of the rope in the fisherman's left hand and the net folded in right hand. All types of fishes are caught with this net.

iii. ***Musharijal:*** This net is extensively used in whole year except monsoon season. In some places like in Kamrup it is also known as '*Mohorijal*'. The net is made up of 2-5 pieces of 20-30 m x 6-8 m, rectangular nylon mosquito nets of mesh size 1-1.2 mm. The pieces are tied together by nylon threads. At the time of netting the net is kept vertically erected from the bottom. It is big size net which required 4-6 persons for fishing.

iv. ***Berjal:*** It is a large rectangular seine net as surrounding net is also known as tanaber and it is similar to mahorijal. The net is shut in a semi circle with the shore as based and is hauled up on to the land by gradually pulling in either end. Two boats each of which carries half of the net do the paying of the net. Operation of this net requires 10-15 person and 2-4 boats. The catch comprises different fishes such as *Labeo* spp., *Cirrhinus* sp., *Mystus* spp., *Channa* spp., *Wallago attu, Ompok* spp. and *Rita rita* etc.

II. Entangling gear

Entangling gear such as gill nets are passive fishing gears, and usually operated by drifting in surface and column layer of water.

i. ***Phansijal***: *Phansijal* is a gill net with rectangular in shape and made up of nylon thread. The operation net involves several indigenous mechanisms to drive the away the fishes into the net. This net is also known as per type of fishes caught *Illisha phansijal, Rou phansijal, Ari phansijal* etc.

ii. ***Langijal* (Gillnet): *Langi jal*** is rectangular net which provided with head and foot edges. This small meshed drift net is more effective in entangling the fish. The size of net is 100 m x 0.5 to 1.5m, bamboo sticks used as floats and clay used as sinkers. The net is tied against the current of water and allowed to drift over night. It is also thrown over the water particularly from one shore to another for fishing. In this category different types net such as *Puthijal, Kaai langi, Gorai langi,* etc. are being used.

III. Drag nets

The drag nets are used for dragging the river beds or large water bodies. They are generally used in shallow water. Drag nets in Assam are large and exist in various forms and dimensions.

i. ***Moijal:*** It is a small trawl net used in shallow waters. The net is rectangular in shape with size 40-50 m X 1.5 – 2 m. A number of small iron drum shaped weights are attached to the lower half of the net. The top of the net is attached with a thin bamboo rod. The net is taken to the deep portion of the river and dragged towards the shore. The catch includes *Puntius* spp., *Mystus* spp., *Clarias* spp. and *Ompok* spp.

ii. ***Harharijal:*** It is a conical shaped, bag like trawl net, 8-10 m length and 6-8 m in circumference at mouth. It is made from cotton or nylon twine. The mouth of net is distended two rods near the two ends where the drag ropes are fastened. The middle of the mouth is kept open by a vertical float of bamboo, attached to the middle of the upper edge. A chain of small drum shaped iron weights in sewn to the lower edge which keeps the net vertical in position. Two small boats are required during operation.

IV. Lift nets

i. ***Dhekijal***: This is a fixed triangular shaped net made up of cotton or nylon. It consist of two long bamboo poles tied together to form an acute angel. The net is kept hanging on a rigid frame in such a manner that it usually put under water by a pulley design. To lift the net, the operator exerts extra pressure on the bass of the triangular corner as weight. This gear is usually operated at the adjoining areas of the *beels*.

ii. ***Parengijal*: *Parengijal* or *Dharmajal***, also called as *Sipjal* is used by all fishermen community of the area, such as *Kaibartas* and also by *Boro* and *Rabha* tribes. This is a mobile dipnet. This square shaped net having area of 4 to 9 m. Each of the four corners is fastened to each of the four ends or two flexible bamboo tied together crosswise at the middle. Another bamboo pole is tied together at the centre of the frame to act as a handle. The net is plunged into the shallow water and raised up at intervals to catch fishes.

Fig. 16: *Parengijal* or *Dharmajal* (*Sip jal*)

Fig. 17: *Dhekijal*

B. Fishing Traps

i. ***Jakoi*:** The *jakoi* is a bamboo trap, which is triangular in shape with a wide mouth (2 m circumstances). It is made of bamboo sieves own at intervals of 0.3-0.7 cm. A bamboo rod is fixed across the mouth from the middle of the base of the triangle to the vertex and prolonged to a short handle. The fisherman plunges it into the bed of shallow water with the mouth facing him and dances on the ground driving the fish at the bottom into the pouch and it is suddenly lifts up in slanting position.

ii. ***Chepa*:** It is drum shaped sieved bamboo trap, tapering at both ends. One opening is closed by a piece of wood while the opening on the other side is half closed by another piece of wood. There is also an opening at the side of the trap with bamboo stripes extended inwards so that fishes get an easy entry to the trap while extended gill like bamboo stripes prevent them from escaping. The trap is facing the opening against the current. The diameter of the opening is 20-40 cm and total length varies from 90 to 150 cm. It is generally used to catch minor group of fishes during the time of April to June.

iii. ***Polo jal*:** The polo jal is a bell shaped slit bamboo trap with a small opening on the top (15 - 25 cm) and bottom (60 - 90 cm). It usually 0.6 - 0.9 m in height and is used in shallow water for catching all kinds of fishes. The fishermen walk into the water press down the polo in front of him and the stopping down plunges his hand through the opening at the top and gropes in the mud for fish that are trapped.

iv. ***Juluki*:** The design detail of Juluki is very much similar to that of polo with few exceptions like the circumference of both the top and bottom opening are almost equal (in mouth 20-30 cm, bottom 28-30 cm). Rice bran bails are used to lure the fishes which facilitated easy trapping.

v. ***Jhupri*:** The design detail is similar to polo but in this trap the mesh size is much smaller which can trap fish fry also. The spaces between the reeds are interwoven with sting. All types of fishes are trapped with *Jhupri*.

vi. ***Ghokajal*:** *Thelajal* or *Ghokajal* is the scooping gear. The net is fitted to a light triangular bamboo frame. The fisherman put their hands in the

Fig. 18: *Jakoi*

Fig. 19: *Khaloi*

Fig. 20: *Polo Trap*

Fig. 21: *Chepa*

Fig. 18-21: Different types of fishing trap made up of bamboo

handle and walks in knee deep water and pushes the free end of the net through the bottom of the river. After a while the net is raised to collect the fishes that may come in the net.

C. Hook and Line Fishing

Here the fishes are individually caught. Though the quantity of catch is less, catch obtained by the line fishing is generally of high quality and of commercial value. The line fishing technique is highly energy efficient, environmentally safe and low cost technology. This fishing is composed of a line and hook. Two types of hooks and line, *Nal barshi, Sip barashi* are widely used for fishing in Assam.

i. ***Nal barshi***: This type of hook bears a *nal* (rod) instead of bamboo, which is of about 1.5 m long and is tied centrally with a nylon rope with a hook, which can float freely at the right angle to the *nal*.

ii. ***Sip barshi*:** It is made up of bamboo, which often exceeds 6m in length. At the tip of the bamboo a nylon rope with a hook is tied with a grasshopper as bait. This type of hook is used to capture fish like *Mystus* and *Clarias*.

3. MIZORAM

The Mizo tribes of north east India developed and practiced different types of fishing methods since time immemorial. The fishing devices are made by using locally available materials and apply indigenous ideas and skills. Some of the fishing gears have been described below:

i. **Poking net *(Hawlhrawk)***: In this method, two sticks are tied at ends of 0 .75 to 1.25m wide net which is about the size of a sieve. It is then pushed between the stones/rocks inside the water bodies. Fishes that are trapped in the net are collected by hand.

ii. **Cast net (*Len den*):** Two types of cast net are used based on the size of river. The fishing net used in big river is called *len-pui* and the smaller one that is used in small river or stream is called len-fâng. The shape of the net is more or less a pyramid shape and prepared from cotton/nylon. About hundreds metallic sinkers are tied at the tip of the net. The net is placed between the two elbows and are thrown to the water bodies.

iii. **Scoop (*Lensuah*):** Such nets are usually operated in the shallow waters for catching small fishes. The net is usually made of cotton or indigenous fiber. This type of fishing gear is used to catch fish for domestic consumption.

iv. **Water diversion (*Thliar leh*):** In this type of fishing a stream is diverted into a small channel by blocking with mud, boulders and plants, so the water from the stream is bailed out and the fishes are caught easily. Most of the fish collected include small sized bottom dwelling fishes like *Garra* sp., *Noemacheillus* sp. and *Mastacembelus* sp.

v. **Temporary water blockage (Suar dan):** In this method both herbal products and temporary water blocking are applied at a shallow place of river. Any available cloths and banana leaves used as fencing to block the running water. Then pool is treated with herbal products like roots of *M. pachycarpa* or bark of *A. procera*. After certain period of time gathered unconcious fishes are collected through small size net (scoop net).

vi. **Pool fencing (*Lui hung*):** A pool where the water is more or less stand or slow running, with depth of about 1-1.5 m in the river or stream is selected. A bamboo fencing having one small door for fish entry is made during day time. Inside this fencing a good smell fish food are spreads as a lure. At night when an adequate number of fishes are attracted inside fencing, the door is closed and fishes are trapped by cast net.

vii. **Cave fencing:** A cave where fishes are abundant is selected and water surrounding the cave is separated by bamboo fencing. The herbal product is used for fishing.

viii. **Pot immersion (*Bel chiah*):** Sometimes smaller fishes used to swim near the bank of a river. A small kitchen pot containing food items is covered with cotton having a small hole at the centre. The pot is then placed inside the water body and fishes are attracted by the lure item. After sometime, fishes enter the pot and thereafter the hole is covered with palm or finger and the pot is taken out.

ix. **Water bailing (*Tui theh kang*):** During dry season, small stream which harbour fishes and prawns are identified. Then the water source is diverted and the water content of the pool is drained in all possible ways. After the water is poured out, then fishes are easily picked up with hand or with any other devices.

x. **Intercept process (*Lui tea sangha lut lo dan*):** Some fishes swim out of main river to smaller tributaries in day time and these fishes return to the river at night. Such fishes are caught by a temporary fencing at a suitable place.

xi. **Barricade (*Bawngdawh*)** this is a bamboo trap of various designs. The principle of the method is to trap the fish in a cage made generally of split bamboo. Sometimes to lure fish foods are kept inside the trap. This method does not need to guard the traps; the fishermen simply put the trap in a specific location and pull out after sometimes.

4. MANIPUR

In Manipur about 45,000 fishermen and fisherwomen fish in the lakes and wetlands of Manipur valley. These fisher-folks who maintain a subsistence economy have no alternative other than using primitive and traditional fishing crafts and gears. These fisher-folks are sinking more and more in this drudgery of working with primitive tools and techniques. The classification of fishing devices of Manipur is shown in the Table 1.

Most of the fishing methods are more or less similar to the general technique practiced in other places of north east, although there are slight modifications in fishing devices & methods to suit the local condition. An account of fishing gears of Manipur is given below.

A. Fishing Without Gear

This is simplest method of catching fish. It needs experience, skill and good knowledge of natural habitats. The common method is "*Khut humba*" (Groping) and used in shallow water. The main species caught are *Chana* sp., *Clarias batrachus, Anabas testudineus, Monopterus albus* (fresh water eel). The strading (*Eashipa*) is another method used for fishing without gears.

B. Fishing with Gears

I. Wounding gear (Spears)

These are implements, which had been used since pre-historic time. The spears are either double pronged or multipronged.

i. **Long** (Spear): The implement locally known as "long" is of wide used. It has 7 to 9 prongs made out of splitting of bamboo provided with iron points at the tips of around 2 inches in length. These prongs are tight to a bamboo pole of 2-3m long.

ii. **Yotchei long:** This differs to common 'long' is that the prongs of Yotchei long are made by iron rods. The gear is used in water bodies including lakes, rivers, reservoir etc. The gear is widely used for the catch of fresh water eel, *Monopterus albus*.

II. Line fishing

The principle of line fishing is to offer bait to attract the fish to bite it. Once the fishes are hooked it prevented to escape fish. The lines are of hand line, set line, pole and line fishing. Basically the line fishing consists of a line and a hook tied to a rod, but it may have 50 to 100 hooks fastened to a long line.

i. **Pole and line fishing:** It is popular fishing techniques in Manipur operated in lake, river, reservoir, irrigation channels, pond etc. A pole measuring about 0.75m and a nylon thread of about the same length with a hook at other end is tied to the tip of the pole. The main species caught by this gear are the *Chana* sp., *Clarias batrachus*, *Mystus gulio* - (Ngasep) etc. Fishing is done throughout the year.

Fig. 22: Pole and line fishing

ii. **Long line fishing:** Long lines may be set stationary or towed through water or left drifting. They set at bottom with the help of weights, or left drifting with the help of floats. It is mainly operate in lakes in winter season. It consists of a main line and a branch line the main line is a nylon twine the length can be of any length according to the convenience generally 150 to 250 m are common. The hooks are tight to this branch line by proper knot. The main line is tied to the available weeds or fixed poles at several points. The common fishes caught by hooks are Ngarins, *Aguilla bengalensis, Mastacembolus armatus* (Ngaril), *Wallago attu, Bagarius bagarius* (Ngaren), *Botia dero.*

Table1. Fishing Techniques of Manipur

A. Fishing without gears		**B. Fishing with gears**					
a) Groping (*Khut humba*)	b) Stranding (*Eeshitpa*)	I) Wounding Gears		II) Line fishing	III) Traps		
		a) Multipronged spear (*Ngarong*)	b) Double/ quadruple pronged spear (*Ngaprum*-long)		a) Tubular traps	b) Spindle	c) Box traps (*Taijep*)
		i) Shoutanba		i) Angling (*Khoi-chopa*)	i) Genuine tubular trap (*Sora-lu*)	i) Single Valve automatic trap (*Kabo-lu*)	i) Taijepthumba
		ii) Mahumtanba		ii) Pole & line (*Khoitek*)	ii) Giant tubular trap (*Tekhao-lu*)	ii) Spindle shaped trap (*Taothum-lu*)	ii) Koirelthumba
		iii) Narengbathinmba		iii) Gorges (*Khoyat*)	iii) Funnel shaped eeltrap (*Lu-lu*)		iii) Phoomda othaba
		iv) Achabathinba		iv) Long line (*Khoisang*)			

B. Fishing with gears							
IV) Nets							V) Stupefying device
a) Dip nets	b) Bag nets	c) Gill nets (*Lang*)	d) Drive-in nets	e) Falling gears		f) Surrounding nets	a) Deoxy genation of water (*Nganeiba*)
				1. Cast Net	2. Cover pots (*Long-up*)		
i) Lift net (*Nupi-een*)	i) Longthrai		i) *Hi-yeiba*	i) Hand thrown cast net (*Nupa-een*)		i) Langoi chatpa	
ii) Rohu net (*Rou-een*)	ii) Thelong		ii) Shamulang thaba	ii) Dragging cast net (*Eenphui*)		ii) Phoom fishing	
iii) Blanket net (*Eenjao*)	iii) Long Longpak		iii) Phaboulang Thaba			iii) Phoom shengba	
						iv) Lukoi touba	v) Koinet touba
						vi) Hapa takapa	

Source: Project Report (2005), Ministry of Science & Techn. Council Manipur, Govt. of Manipur.

III. Trap

In trap, the fish enters voluntarily but it hampered from coming out. The following are some of the traps used in Manipur:

i. ***Taijep*:** This trap is rectangular in shape all the six sides of the structure surrounded by bamboo netting. An entrance is provided at the middle of one side of the longitudinal portion where the fish to enter, but prevent to escape by fixing a comb like bamboo strips which are fitted at a rubber string which is of flexible commonly known as funnel. Once the fish is entered it is trapped inside.

Fig. 23 *Taijep* (Source: Anon., 2005)

The material used for the construction of *Taijep* is bamboo strips of small and finger size; this is tight by a synthetic yarn. The size of trap varies in length and in breadth and having funnel shape mouth. This traps are set longitudinally by tightening to a floating weeds or by submerging 3/4 portion to the gaps found at floating weeds. They are also operated in shallow water near the shore of channels, paddy fields etc. either single or even in series. The main species caught by this gear include *Notopterus notopterus, Channa* sp., *Anabas testudeneous, Clarias batrachus,* Common carp etc.

ii. ***Tekhao-lu*:** The gear is a tubular shape tapering at the end, made by the bamboo splits. A round valve with inwardly projected pointed bamboo sticks is fixed at the mouth of the trap so as to prevent the escapement of fishes commonly known as "sou". There are various sizes of *takhou-lu.*

Bamboo strips are fixed to these bamboo rings at a gap of one inch, which is less at the cod end. These bamboo rings and strips are tight together by the nylon yarns. The season of operation is from rainy and winter season. They are operated only in the running water. One bamboo pole or wooden pole is fixed at the bottom of the riverbed at the depth where the gear is to be lower; with the help of one rope the cod end is tight. Two another ropes is also tight to the gear near the mouth opening. After lowering the gear at required depth these two ropes are fixed to the shore pole to keep the gear in position. The gear mouth is keep along the direction of the river flow. The fish come up against the current (i.e., migratory fishes) are trap inside. This trap (*Takhou lu*) is suppose to be the largest trap operates in Manipur.

iii. ***Kao*:** This gear is a peculiar indigenous trap used in the river system. It is rectangular in shape having a length of 1.5 m and height of 0.75 m, breadth of about 0.9 m at the mouth region. This gear is made by interweaving of bamboo splits with spaces sufficiently to pass the water. One end of the bamboo netting is fold lengthwise in the middle to join into a single vertical line whereas the other end is left freely open. Bamboo frames are provided all around its mouth. The setting of the gear is done along with the current. Two bamboo poles are fixed in which the trap is keep intake to avoid drawing by the river current. In some cases the mouth is keep opposite to the current. Green branches of trees and bamboo are placed inside the trap. It acts as shelter/hiding place for the fishes. The upper portion of the trap is fully open. Sometimes rice bran or some bait is keep inside to attract the fish. It is usually lifted after an interval of 4 to 5 hours. The main species of fish caught by this gear are the bottom dwelling fishes commonly *Mystus* sp., *Clarius batrachus, Chana* sp. *H. fossilis*,

iv. ***Kabo-Lu*:** It is an oval shape trap made by the small bamboo splits. There are various sizes ranging from two to four (0.5 m to 1.25 m) feet in length having a middle diameter half to one foot (½ to1 ft) or even more according to the requirement. The one end of the trap provides a neck with a mouth from which catches can be taken out. The diameter of the mouth vary according to the size of the trap ranging from 5 to 10 cm. A funnel is provided near the middle portion with certain pointed bamboo splits. According to the ascending and descending behaviour of the fishes. These traps are fitted to the bund.

v. ***Lushat* (*Soralu*):** This gear is very much similar to that of the *Tekhoulu*, having a tubular shape made by the bamboo splits, but without a funnel unlike in the *Tekhoulu*.

IV. Nets

a) Dip net (Lift net)

Here the net is rectangular in shape. The net is dip after an interval and again lift. The fishes come voluntarily sometimes some bait such as rice bran is broadcast over the net to attract the fish. Depending on the size & mode of operation they are of two types:

i. ***Nupi-il* (Lift net):** Two bamboo frames are tied at the middle by a rope thus divided into 4 tips. The nets are fixed at these 4 tips forming the structure as cross mark "X". A bamboo pole is fixed at the middle of the cross; this will act as lever to lift the net. This bamboo pole is commonly known as *Poura*. In some cases one rope is attached to this pole to lift the net.

Fig. 24: Lift net (*Nupi-il*) in the Loktak Lake

The lift net (*Nupi-il*) operated in the Loktak Lake and its surrounding peripheral area is of rope less lift net. This gear is one of the commonly used in Manipur by the fisher women hence the name "*Nupi-il*". In the ropeless *nupi-il* the *poura* (bamboo pole) is inserted between the thighs and pressing it drawn, thus the net is lift. The catch by this gear is a mixed species.

ii. ***Injao*:** The larger size of lift net is called "*Injao*". In this, the rectangular net is made by joining 9 pieces of net. The bamboo structure fitted at four corners and net is tightened. If operation is done in open water-craft is used to lift the net as an operating place or a floating weed mass commonly known as "*phum*" is used as platform. The lowering or setting of the net is done by lowering the rope provided at one side, the lifting of the net is done by pulling the two ropes at one end. This type of *Injao* operation is done only in the lakes. The catches are of mixed species. It is operated throughout the year. The peak period is from September to March.

iii. ***Lang-il*:** This is a peculiar type of lift net seen in the Kongba river near the Kongba Irong. The difference of this net from that of the small dip net (*nupi-il*) is that in the case of *lang-il* a bamboo structure is provided as a working place. A bamboo rack is installed at the riverbank with the help of bamboo. The size of the net is also comparatively larger having a rectangular shape of 4.5 m in length and 4 m in width. The lifting and lowering of the net is done by pulling and releasing the rope fixed to a bamboo pole which in turn fixed to the cross "X" shaped bamboo split structure where the net is fitted. The operation is done at the platform provided at the bamboo rack. The species caught by this net are of larger size fishes. In this particular area the operation of this nets are done when the Imphal river water flow into the Kongba river. The main season is during the month of June to September. The main species caught are *Wallago attu*, *Osteobrama belangeri*, *Labeo dero* (*Khabak*), *Labeo angra*, (*Ngaton*), Common carp, *Labeo rohita*, *Heteropneustes fossilis*, Indian major carps, *Anguilla bengalensis* (*Ngarin leina*) etc.

b. Scoop net

Here the net is fixed to a frame, it can be of fully fixing or fixed when required. The mesh size of the net varies according to the type of fishes expected to be caught. Here the net move with the help of manpower, it can be operated in almost all the water body. Usually a handle is provided to hold the net in proper shape by the fisherman. There are small scoop net without any handle, they are mainly operated at the shallow water only. The most commonly used scoop nets are:

i. ***Longthrai***: This gear is operated either from shore or from a canoe. This gear consists of an elliptical shape bamboo ring, which is fixed to a bamboo pole. The circumference of the ring is around 5m. The net is fixed to this elliptical bamboo ring. The shape of the net is cone shaped tapering at the bottom. The most commonly used mesh size of this gear is the number 20 net. Few large size mesh is also found in use for catching large fish. The bamboo handle is around 5 m. The gear is used almost throughout the year at different depth and different water bodies. The main species caught are *Punctius chola* (*Phounga*), *Channa* sp., *Cyprinus carpio, Notopterus notopterus*.

Fig. 25: *Longthrai*

Fig. 26: Fishermen operating of Long for fishing

ii. ***Long:*** In a bamboo ring of diameter around 0.6 m, the bamboo-split netting in the form of a half-elliptical shape is fitted. The mesh size is quite small just as to filter the water. This operates only in the shallow water, mainly small channel paddy field and near the shallow water of river. This gear is one of the indigenous gears used in Manipur and operated by females for catching small fishes, shrimps etc.

iii. ***Mu* (*Mu-namba*):** This is one of the indigenous gears used by fishermen in the riverine fisheries. Commonly used at the Imphal river system. The gear consists of two bamboo poles used as handle and fixed the net on it at the tip ends. The bamboo poles have a length of 5 m in which a bag net is fitted. The mouth portion is fitted to this bamboo poles. The length of the net is around 4m in length. Depth of the net is around 3m. The net is operated in the running water only to catch the migratory fishes. The net is operated by two fishermen; one fisherman operates the net by dipping the net completely in the water by opening the net mouth with the help of the bamboo poles. The opening of the mouth is along with the current, whereas the other man rows the canoe. Migrating fishes that ascend against the current for breeding or in search of food are entangled in the netting. As the fisherman feels the entanglement of the fish, he clips the net mouth, lift it up and take out the fish. The main season of this gear is the June to August and November to February. The main species caught by this net are I.M.C. (Indian Major Carps), *Labeo angra* (*Ngaton*), *Botia dero, Labeo dero* (*Khabak*), *Bagarius bagarius* (*Ngaren*), *Wallago attu* and other fishes that are of migratory habit.

c. Gill net

These are single walled nets with a mesh opening of such a size that the required fish can gill themselves in the netting. This is a passive gear, but fish can also be driven into the gill nets. The nets are used singly or in series. This is the most widely used fishing gear. They are vertical walls of netting normally set out in a straight line. Floats and sinkers are attached plastic ropes are used as head ropes and foot ropes. There are various mesh size of gill net varies almost 0 to 12cm. The gill nets are directly purchase from the market or purchase as twine and making of net is done at home. The other materials like footrope, head rope, float, sinker etc. are also purchased from the market.

d. Falling gears

i. **Cast Net:** These nets are commonly known as "*Nupa-il*". These are conical shape nets. The lower edge is provided with footrope attaching a series of sinkers to it. This is a falling gear; here the fishing is done by direct quick movement of the gear to cover the fish. It is mainly operate in shallow water. There are two types of cast net

ii. **"*Nupa-il*" (*Shangbran*):** it is a small size cast net operation is done by a single person the fisherman twist the body and hand through the net over the fish shoal when expected. Due to the sinker provided the gear sinks quickly in the form of an umbrella thus cover the fish. This gear is one of the common gears used in Manipur.

Fig. 27: *Nupa-il* (Cast net)

iii. **"*Inphul*":** The bigger size cast net is commonly known as "*Inphul*". The operation of this gear is done by dragging 2 or 3 persons required for the operation. Half of the net is drawn at the bottom of the fishing ground the upper portion of the net is hold by hand after a short distance the upper portion is lowered again. The hauling of the gear is done by pulling the string provided at the cone end. The catch is a mixed species. The smaller size cost around Rs. 750/- to Rs. 1000/-. The larger size *Inphul* cost around Rs. 2500/- to Rs. 3000/- depending on the size.

e. Drive in nets

This group contains all nets, begs or net walls which are towed through the water on or near the bottom. The manner of capturing is filtering the passive prey by the active moving gear. The net is sometimes made by stitching many pieces and provided with floats and sinkers fastened to the head and foot ropes, according to the depth of the water body and area covered size of net is made, this may be temporary stitch depending on size i.e., the depth and the length, suitable float and sinkers are fitted. According to the size of the fish expected to be caught the mesh size of the net is selected.

i. ***Phabou lang:*** The name of *Phabou lang is* because of main species caught by these nets are of *Puntius* spp. (*Puntius chola* - Phabounga). The other important fish caught in Manipur by these nets are the *Notopterus notopterus* and other similar size fishes. This net operates throughout the year in the lakes of Manipur.

f. Surrounding net

The mode of capture is done by surrounding not only from the side but also from the beneath, thus permitting the capture of fish over very deep waters by preventing their escape into the depths.

i. ***Phum fishing:*** This fishing method is extensively practiced in Loktak Lake, is a characteristic feature of Loktak lake. Number of Athaphums (Phum circles for fishing purpose) has increased considerably after the construction of Ithai barrage. As per remote sensing imageries, the number of Athaphums has increased from 217 in 1990 to 3019 in February 1999. At present, there are about 3200 phum circles in Loktak lake covering an area of about 10.5sq. km. Fish production from *athaphum* (*Phum*) contributes to about 39% of total capture fishery yield of the lake (558m).

ii. **Phumdis:** *Phumdi* are a series of floating islands, exclusive to the Loktak Lake in Manipur state, in northeastern India.

iii. ***Phum thaba*:** The floating mass of weeds commonly known as the *phum* is cut from the wild mass. These pieces of *phum* are transplanted in a circular form in s suitable water surface at the lake, which serves as shelter for wide varieties of fishes. This *phum* is fixed with the help of a bamboo stakes in a circular form. The total length of the *phum* vary ranging from 150 to 200 m. The width of the *phum* is around 3m. After setting this floating mass in circular shape fishes are allow taking shelter,

Fig. 28 & 29: *Phum* – floating mass of weeds at Loktak lake

in most cases to attract the fishes rice bran are spread inside this circle. After keeping this for a few days or in months. According to the assumption by the expert fisherman of expecting good gathering of fishes.

iv. ***Phum namba*:** Here from the inner side of the ring the net is release to encircle the area weight is provided at the footrope. The upper portion of the net is fixed at the floating mass. The purpose of this is to prevent the fish to escape. Now from the inner side as well as from the outer side are totally blocked by this net. This net is of very small mesh size in olden days tarpaulin was used. Thus enclosed water area is stirred from the bottom with the help of bamboo poles by a group of fishermen (15 to 20). This process is commonly known as "*leihotpa*". This cause the turbid to the body and thus suffocate the fishes, thus the fishes come up to the surface water. During this time the fishes are caught by using a scoop net (*Longthrai*). There are two main season for this fisheries. September to October again at February to March. The catches are of mixed species. *Notopterus notopterus, Cyprinus carpio, Hypophthalmichthys molitrix, Amblypharyngodon mola, Cirrhinus mrigala, Puntius* sp. *Channa* sp., *Labio rohita, Clarias batrachus, Ctenopharyngodon idella,* etc. The main important fishing activities carried out in the hill streams of Manipur includes:

CONCLUSION

In this documentation of the traditional fishing practices and fishing devices of four states such as Tripura Assam, Mizoram and Manipur, it is observed that the North Eastern states are not only rich in natural resources, biodiversity particularly fish biodiversity, but also rich traditional knowledge of fishing. There are wide ranges of fishing gears and craft are reported from the region. The gears which include hooks, spears, traps and nets. The fabrication shape size and design are varied from one state to other state as they developed through generation based on experience and as per requirements. This rich traditional knowledge of the people in fishing is further needed to be studied and documented for preservation and further developments in fishing technologies for the wellbeing of fishermen communities of a region.

References & Recommended Readings

Anon., 2005. Survey and modernization of the traditional Fishing crafts and gears in the lakes and Wetlands of Manipur. Project Report, Manipur Sc. & Tech. Council Central Jail Road, Imphal. pp.65.

Blackwell Science Ltd., Oxford. 368pp.

Brandt, A. V., 1984. Fish catching Methods for the World (3rd ed.). Fishing News Books Ltd., London, pp. 418.

Chakravartty, P. and Sharma, S., 2013. Different Types of Fishing Gears Used by the Fishermen in Nalbari District of Assam. *Int. J. of Soc. Sc. & Interdisciplinary Res.*, **2**(3): 177-191.

Chatia, R. I. and Seydur R., 2007. Traditional Riverine Fish Catching Devices of Assam. *Fishery Technology* **44**(2): 137-146.

Devi, B. N., Mishra, S. K., Das, L., Pawar, N. A., Chanu, Th. I., 2013. *Traditional fishing methods* in Central valley region of Manipur, India, *Indian Journal of Traditional Knowledge,* **12**(1): 137-143.

Devi, B. N., Mishra, S. K., Pawar, N. A., Das, L., Das, S., 2013. Traditional fish aggregating wisdom of Manipur, Northeastern India. *Indian Journal of Traditional Knowledge* **12**(1): 130-136.

Dutta, R. and Bhattacharjya, B. K., 2009. Traditional Fishing method of Assam for Catfishes using duck meat as an attractant. *Indian Journal of Traditional Knowledge* **8**(2): 234-236

fishing using wind energy along with Tamilnadu coast. *Proc. Nat Workshop on Low Energy Fishing Soc. of Fish Technol. India* 272-277.

George, V. C., 1971. Fishing techniques of riverine and reservoir systems: Present status and future challenges, Riverine and Reservoir fisheries, SOFT(I), 192-196.

Hickling, C. F., 1961. Tropical inland fisheries Publication. Longman, Green and Co. Ltd. London, 287 pp.

Hornell, J., 1923. The fishing methods of the Ganges. *Mem. Asiatic Soc. Bengal* **8**(3): 199-237.

Hornell, J., 1938. Fishing Methods of Madras Presidency, Part II: The Malabar Coast, *Madras Fish. Bull.* **27**(1): 69.

Islam M. R., Das, B., Baruah, D., Biswas, S. P. and Gupta, A., 2013. Fish Diversity and Fishing Gears used in the Kulsi River of Assam, India. *Annals of Biological Research* **4**(1): 289-293.

Jhingran, A. G. and Natarajan, A. V., 1969. Study of the fisheries and fish population of the Chilka lake during the period 1957-65. *J. Inland Fish. Assoc. India* **1**: 49-63.

Job, T. J. and Pantulu, V. R., 1953. Fish trapping in India. *J. Asiatic. Soc. Sci.* **19**(2): 175-196.

Jones, S. and Sujansingani, K. H., 1952. Notes on the crab fishery of Chilka lake. *J. Bombay Nat. Hist. Soc.* **51**(1): 119-34.

Kar, D. and Dey, S. C., 1993. Variegated encircling gears in Lake Sone of Assam. *J. of Appl. Zool. Res.* **4**(2): 171-175.

Lalthanzara, H. and Lalthanpuii, P. B., 2009. Traditional Fishing Methods in Rivers and Streams of Mizoram, North-East India. *Research Note Sci. Vis.* **9**(4): 188-194.

Mitra, P. M., Ghosh, K. K., Saigal, B. N., Sarkar, N. D., Roy, A. K., Mondal, N. C., and Paul, A. R., 1987. Fishing gears in the upper and middle Hooghly estuary. Bullein. Central Inland Fisheries Research Institute, Barrackpore **49**:1-22.

Miyamoto, H., 1962. A field manual suggested for fishing gear surveys (In Mimeo), CIFT, Cochin, 15 pp.

Mohanrajan, M., 1993. Fish trapping devices and methods of southern India, *Fish. Technol.* **36**: 85-92.

Nair, P. R., 1993. Fishing with traps. In: Low Energy Fishing. Fish. Technol. Special Issue) Society of Fisheries Technologists (India), Cochin, p. 207-209.

Nath, P. and Dey, S. C., 1989. Fish and fisheries of Northeastern India Vol. 1 (Arunachal Pradesh), 194 pp.

Prabhu, M. S., 1954. Trap fishing. *Indian J. Fish.* **1**: 94-129.

Pravin, P. and Meenakumari, B., 2008. Fishing traps of Assam, *CIFT Special Bull.* No. 14, 162 pp.

Pravin, P., Meenakumari, B., Baiju, M., Barman, J., Baruah, D. and Kakati, B., 2011. Fish trapping devices and methods in Assam - A review. *Indian J. Fish.* **58**(2): 127-135, 2011

Remesan, M. P., Pravin, P. and Meenakumari, B. 2007. Collapsible fish traps developed for inland fishing. *ICAR News.* **13**(1):117-118.

Sainsbury, J. C., 1996. Commercial Fishing Methods (an introduction to vessel and gear).

Sathinshas, R., Panikkar, K. K. P., and Salini, K. P., 1993. Economic of traditional gill net.

Sharma, P. and Ahamed, S., 1998. Relative efficiency of fish capturing devices in Kachodhara Beel of Morigaon district, Assam. *Environ. Ecol.* **16**(1):123-126.

Sharma, P., Kalita, K. K. and Dutta, O.K., 1993. Low energy fishing techniques of the North eastern India. In: George, V. C., Vijayan, V., Varghese, M. D., Radhalakshmi, K., Thomas, S.N. and Jose Joseph (Eds.), *Low Energy Fishing, Fish. Technol.,* Society of Fisheries Technologist (India), Coch. p. 163-167.

Sharma, R., 2001. Traditional fishing methods and fishing gears of Assam. *Fishing Chimes* **20**(12): 23-26.

Slack-Smith, R. J., 2001. Fishing with traps and pots. FAO training series 26: 66 pp.

Thomas, S.N. Edappazham, G., Meenakumari, B. and Ashraf, P. M., 2007. Fishing Hooks: A Overview. *Fishery Technology* **44**(1): 1-16.

Tynsong, H. and Tiwari B. K., 2008. The traditional knowledge associated with fish harvesting practices of War Khasi community of Meghalaya. *Indian Journal of Traditional Knowledge* **7**(4): 618-623.

Upadhyay, A.D., Singh, B.K., 2013. Indigenous fishing devices in use of capture fishing in Tripura. *Indian Journal of Traditional Knowledge* **12**(1): 149-156.

Varghese, M. D., George, V. C., Gopalkrishna Pillai, A. G and Radhalakshami, K., 1997 Properties and Performance of fishing hooks. *Fishery Technology* **34**: 39-44.

Verma, A. M., 2007. Fishing Gears used by Riparian Fisheries posing threat to the conservation of rare fish fauna in North Bihar. *Fishery Technology* **44**(2): 147-152.

Wilimovsky, N .J. and Alverson, D. L., 1971. In: Kristjonsson, H. (Ed.), *Modern Fishing Gear of the World,* Vol. 3, Fishing News (Books) Ltd., London, 509 pp.

Yadava, Y. S. and Choudhury, M., 1986. Banas fishing in beels of Assam. *J. B. Hist. Soc.* **83**(2): 452.

Yadava, Y. S., Choudhury, M. and Kolekar, V., 1981. Fishing methods of flood plain lakes in North Eastern region. *J. Inland Fish. Soc. India* **13**(1): 82-86.

Chapter – 22

Use of Black Drongo for Pest Management in Rice: A Traditional Agricultural Practice by the Rural Farmers of Tripura

Partha Pratim Bhattacharjee, Joydeb Majumder and Pranabjit Bardhan Roy

INTRODUCTION

North East India constitutes the states of Arunachal Pradesh, Assam, Manipur, Meghalaya, Mizoram, Nagaland, Tripura and Sikkim which reflects ecological and cultural contrasts between the hills and the plains. It is one of the most biodiversity regions in the world, with more than a hundred and fifty tribes forming a melting pot of variegated cultural mosaic of people and races constitute a vital component in this region (Dutta and Dutta, 2005).

More than 70% people of the region are engaged in agriculture while rice is the staple food crop (Bhattacharjee and Ray, 2010). The farmers from different ethnic diversity of the region have developed their own systems of cultivation of various crops (Islam et al. 2004). These indigenous traditional knowledge systems are quite endemic to this region (Deka et al. 2006). Traditional farmers of this region use various plants and animals for sustainable pest management of different crops, mainly in paddy cultivation. Rice crop has relatively a large number of insect pests especially in tropical regions. Since the crop is attacked by pests right from the time of sowing till it is harvested. The present paper describes the use of black drongo, a highly insectivorous bird for the management of insect pests of paddy.

The Black Drongo, *Dicrurus macrocercus* (Vieillot, 1817) is a small Asian passerine bird of the family Dicruridae. It is a common resident in much of tropical southern Asia and distributed from southwest Iran through India and Sri Lanka east to southern China and Indonesia. They become active very early at dawn and roost later than many other birds. They are commonly found in the rice environments of South Asia and feed mainly on insects such as grasshoppers, cicadas, termites, wasps, bees, ants, moths, beetles and dragonflies (Fischer, 1923). In Tripura, black drongo is a most common bird (Bhattacharjee et al., 2013) distributed throughout the state, mainly in and around the agricultural fields.

Use of black drongo in pest management is a year old practice in this region which is eco-friendly and compatible to other pest management practices. Further more, it helps in conservation of the biodiversity and don't have any detrimental effect on ecosystem.

MATERIALS AND METHODS

The study was conducted in Ichanchandranagar, two km away from Tripura University campus. The geographical location of the area is 23°46′00.37" N and 91°14′54.24" E with an altitude of 17m above sea level. Vegetation structure of the region comprises of agricultural lands surrounded by fragmented secondary forest patches dominated by *Albizia* spp., *Bamboo* spp., *Holarrhena antidysenterica*, *Cassia* spp., *Microcos paniculata*, Rubber plantations, home garden plantations etc. and secondary successional vegetations dominated by *Melastoma* sp., *Chromolaena odorata*, *Clerodendrum* sp., cane, *Imperata* sp. etc. Paddy is the principal crop of the region and vegetables are also grown seasonally (Fig. 1).

Twenty five farmers of Ichanchandranagar village were interviewed during 2011-12 for confirming their traditional agricultural practices in paddy cultivation. Their views regarding the use of black drongo bird in pest management were noted and digitally recorded in different crop seasons.

Fig.1: Satellite view of the study area (Ichanchandranagar) showing the fragmented forest patches with paddy fields

RESULTS AND DISCUSSION

Farmers of this region usually place bamboo sticks or wooded perches in paddy fields to attract the insect-predatory birds like black drongo. These perches serve as watch towers and resting places for these birds during the hunting stage (Fig. 2 & 3). Observations and farmers' view revealed that mainly small holding farmers regularly practice this indigenous pest management. However, medium and large holding farmers often practice modern pest management technologies with chemical pesticides.

Traditional farmers of Assam also places bamboo sticks in paddy field to attract birds, to take rest on it and for controlling stem borer and swarming caterpillars (Deka et al., 2006).

Fig. 2: A single black drongo bird in a paddy field watching for insects

Fig. 3 Placing of bamboo sticks to settle black drongos in the fields

CONCLUSION

Nowadays, the world's indigenous knowledge base is as much at risk as is the indigenous flora and fauna, which should be preserved to enhance the quality of life of those who have generated it. Agro-ecological and ethno-ecological evidence increasingly indicates that these systems are productive, sustainable, ecologically sound, and tuned to the social, economic and cultural features of any farmers.

ACKNOWLEDGEMENT

Authors are thankful to the local people for sharing their ancestral knowledge. Special thanks to Mr. Dipankar Kishore Sinha for his helping hand during field study.

References & Recommended Readings

Bhattacharjee, P. P. and Ray, D. C., 2010. Pest management beliefs and practices of *Manipuri* rice farmers in Barak Valley, Assam. *Indian Journal of Traditional Knowledge* **9**(4): 673-676.

Bhattacharjee, P. P., Lodh, R., Laskar, D., Majumder, J. and Agarwala, B. K., 2013. An ornithological survey in the vicinity of Agartala city of Tripura state, north-eastern India. *Journal of Research in Biology* **3**(3): 852-860.

Deka, M. K., Bhuyan, M. and Hazarika, L. K., 2006. Traditional pest management practices of Assam, *Indian Journal of Traditional Knowledge* **5**(1):75-78.

Dutta, B. K. and Dutta, P. K., 2005. Potential of ethnobotanical studies in North East India: An overview, *Indian Journal of Traditional Knowledge* **4**(1): 7-14.

Fischer, C. E. C., 1923. "Drongo and cicada". *Journal of Bombay Natural History Society* **29**(3): 839–840.

Islam, Z., Heong, K. L., Bell, M., Hazarika, L.K., Rajhkowa, A.J., Ali, S., Dutta, B.C. and Bhuyan, M., 2004. Current status of rice pests and their management in Assam, India- a discussion with extension agents, *International Rice Research Newsletter* **29**(2): 95-97.

ACKNOWLEDGEMENT

[illegible]

Chapter – 23

On Some Aspects of Sampling Methodology with Reference to Indigenous Traditional Knowledge

Prasenjit Pal

INTRODUCTION

Traditional knowledge is the knowledge, innovations and practices of indigenous and local communities around the world. Developed from experience gained over centuries and adapted to the local culture and environment, it is transmitted orally from generation to generation. It tends to be collectively owned and takes the form of stories, songs, folklore, proverbs, cultural values, beliefs, rituals, community laws, local language. Sometimes it is referred to as an oral tradition for it is practiced, sung, danced, painted, carved, chanted and performed down through millennia. Traditional knowledge is mainly of a practical nature, particularly in such fields as fisheries, health, horticulture, forestry and environmental management in general. Appreciation of the value of traditional knowledge is growing. This knowledge is valuable not only to those who depend on it in their daily lives, but to modern industry and farming as well. Many widely used products, such as plant-based medicines, health products and cosmetics, are derived from traditional knowledge. Other such valuable products include agricultural and non-wood forest products as well as handicraft. Traditional knowledge can make a significant contribution to sustainable development.

Most indigenous and local communities are situated in areas where the vast majority of the world's genetic resources are found. Many of them have cultivated and used biodiversity in a sustainable way for thousands of years. Some of their practices have been proven to enhance and promote biodiversity at the local level and aid in maintaining healthy ecosystems. However, the contribution of indigenous and local communities to the conservation and sustainable use of biodiversity goes far beyond their role as natural resource managers. Their skills and techniques provide valuable information to the global community and a useful model for biodiversity policies. Furthermore, as on-site communities with extensive knowledge of local environments, indigenous and local communities are most directly involved with conservation and sustainable use.

The indigenous and local communities have developed and maintained some local ecological knowledge and practices that can have significant implications in scientific studies and on the management of natural/fisheries resources. As for instance, the practices of restricted cutting of *Typha*, fishing and access in sacred sites and conservation of *mabawe.* These traditional practices encouraged regeneration and sustainable utilisation of fish. The knowledge systems have been conserved and passed on from generation to generation through religious beliefs, taboos and myths. Some indigenous knowledge systems have been eroded over the past years due to changes in social structures, immigration and advent of new religions, adoption of new resource harvesting techniques and changes in life styles. Although these knowledge systems are not specifically meant for conservation of natural resources, the study argues that to achieve sustainable designs or implementation of natural resource management projects, there is need to integrate relevant existing indigenous knowledge systems that promote conservation of resources. Johnson (1992) defines IK as a body of knowledge built up by a group of people through generations of living in close contact with nature. According to Johnson, this knowledge includes a system of classification, a set of empirical observations about the local environment and a system of self-management that governs resource use. This involves an ability to analyse situations and interpret the results; practical experience of tackling problems and perspective of having an overview of a subject or issue. This may mean skill and knowledge of what is good to collect or eat, where best to find particular resources and when and how to avoid resource degradation. Thus, ITK builds upon the experiences of the past and adapts to present changes in technology and socio-economics.

SAMPLING METHODOLOGY

To deal with the knowledge/information in the form of stories, songs, folklore, proverbs, cultural values, beliefs, rituals, community laws obtained from indigenous and local communities need to be collected and assessed properly to conclude a valid conclusion from it. An appropriate sampling methodology (like Systematic, Snowball sampling) can serve the purpose. Systematic sampling employs a simple rule of selecting every k^{th} unit starting with a number chosen at random from 1 to k as the random start. Let us assume that N sampling units in the population are numbered 1 to N. To select a systematic sample of n units, we take a unit at random from the first k units and then every k^{th} sampling unit is selected to form the sample. The constant k is known as the *sampling interval* and is taken as the integer nearest to N/n, the inverse of the sampling fraction.

PARAMETER ESTIMATION

The estimate for the population mean per unit is given by the sample mean

$$\hat{\bar{Y}} = \bar{y} = \frac{\sum_{i=1}^{n} y_i}{n}$$

where n is the number of units in the sample.

In the case of systematic strip surveys or, in general, any one dimensional systematic sampling, an approximation to the standard error may be obtained from the differences between pairs of successive units. If there are n units enumerated in the systematic sample, there will be (n-1) differences. The variance per unit is therefore, given by the sum of squares of the differences divided by twice the number of differences. Thus, if $y_1, y_2, \ldots, y_n$ are the observed values (say volume) for the n units in the systematic sample and defining the first difference $d(y_i)$ as given below,

$d(y_i) = y_{(i+1)} - y_{(i)}$; ($i$ = 1, 2,..., n -1), the approximate variance per unit is estimated as

$$\hat{V}(\hat{\bar{Y}}) = \frac{1}{2n(n-1)} \sum_{i=1}^{n-1} \left[d(y_i)\right]^2$$

Suppose, a sample of 200 households was taken from all the 13 villages found on an particular area. The area has 600 households most of which are semi permanent and others are traditional or temporary. The sample selected for the study may not represent the whole country, because the study is only concentrated in a small area. Furthermore, ITK is site specific. Hence, this sample only represented the population of that area where the study was undertaken.

SAMPLING OF THE STUDY POPULATION

1. Systematic Sampling of the Households

The sampling frame to be obtained from the National Statistics Office lists of households for the current Population and Household Census. Say, the sampling frame for the area has 600 households on its list. A systematically selected sample of 200 households to be drawn from this frame. This sample reflected the proportion of the total population in each village. The sample to be selected at an interval based on the following formula:

$$k= N/n$$

Where k is the sampling interval, N is the total number of households and n is the sample size (Burns, 1994). Say, n=200, N=600 and sampling interval k=600/200=3.Since k is 3 therefore every third household from the list was selected. The first household was randomly selected from the first three households on the list. A table of random numbers was used to decide on the starting point. If the kth household no longer existed in the village, the next household in the line $(k+1)$ was included in the sample. After this occurrence, the rest of the households in the sample shifted automatically. In the selected households, all household heads or their spouses were targeted for the survey to capture the views of the middle aged and elderly population.

2. Snowball Rolling for Sampling the Key Informants

Snowball sampling (or chain sampling, chain-referral sampling, referral sampling) is a non-probability sampling technique where existing study subjects recruit future subjects from among their acquaintances. Thus, the sample group appears to grow like a rolling snowball. As the sample builds up, enough data is gathered to be useful for research. This sampling technique is often used in hidden populations which are difficult for researchers to access.

Snowball rolling to be applied to identify key informants interviewed for specific information. The traditional authority to be used to identify one respondent who can help to identify and locate another key informant. This process continued until 20 key informants to be sampled. The key informants to be interviewed to supplement and provide in depth explanation on the information obtained from the individual interviews and the focus group discussions (FGDs). The key informants included the Traditional Authority, chiefs, fisheries officer, the oldest inhabitants who included all inhabitants who were aged 60 and above years, and other people who reportedly, more knowledgeable of the traditional practices and beliefs associated with a variety of resources.

DATA COLLECTION TECHNIQUES

Within all the villages the first objective is to identify different natural resources as defined by the local people. The second objective involves gathering indigenous information on the perceived habitats of these resources. Local people to be asked to describe the practices that exist(ed) on the area and their effect on fisheries management. Fisheries management to be targeted because people's livelihood is highly dependent on fishing and thus the local body of knowledge to be more detailed and accessible than the other resources. A range of techniques to be used to investigate the existing knowledge and traditional management systems (TMS): lets take an examples of 200 household questionnaires, four FGDs and 20 key informants' were interviewed. Say, about 56% of the household respondents interviewed are male and 44% are female. The issues that to be addressed in the household interviews should be discussed in detail in the FGDs. The household interviews to be conducted before the FGDs so that information gathered would be individual and reliable for comparison with information obtained from the FGDs. The FGDs to be used because the participants' thought to be more knowledgeable of the discussed issues compared to other household respondents. The FGD participants included the indigenous inhabitants and those migrants who had stayed on the Island for more than 10 years and had vast ecological and TMS knowledge. The Traditional authority if it is there may render help to identify the members. Selection of the FGDs participants to be done based on their recognised status as fishing experts. The FGDs comprised 10 to 12 participants. Four FGDs, one from each of the social groups (males, females, the oldest members and the youth), to be conducted on four separate days due to cultural constraints. Rural women can express their views in meetings combined with men. Care should be taken to make sure that no one dominated the discussions.

This combination contributes to a thorough data collection because one method cannot provide all the needed information about the respondents. The responses obtained from various data collection techniques need to be compared to relevant information documented in the literature. Quantitative data from the household interviews to be collected and further to be analysed. Descriptive statistics to be worked out to create frequency tables and percentages showing the proportions of respondents who have knowledge of natural resources, their habits and particular traditional management practices. All qualitative data obtained from the FGDs to be compiled and transcribed using line by line coding.

The coding involved categorizing all emerging issues into different themes. This analysis to be applied to list and describe the resources and practices reported on the area.

References & Recommended Readings

Burns, R., 1994. *Introduction to Research Methods*, 2nd ed. Longman, Melbourne.

Goodman, L. A., 1961. Snowball sampling. *Annals of Mathematical Statistics* **32** (1): 148–170.

Johnson, M. (ed), 1992. *Lore: Capturing Traditional Environmental Knowledge*, IDRC,Ottawa.

Ken, B., 2004. *Business Statistics for Contemporary Decision Making* (4th Wiley Student Edition for India). Wiley-India. ISBN 978-81-265-0809-9.

Salganik, M. J. and Heckathorn, D. D., 2004. Sampling and Estimation in Hidden Populations Using Respondent-Driven Sampling. *Sociological Methodology* **34** (1): 193–239.

FSC
www.fsc.org
MIX
Papier | Fördert gute Waldnutzung
FSC® C083411

Zeitfracht Medien GmbH
Ferdinand-Jühlke-Straße 7
99095 Erfurt, Deutschland
produktsicherheit@kolibri360.de